MORE EVERYDAY ENGINEERING

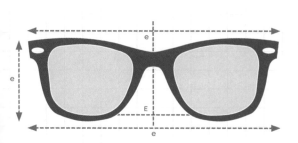

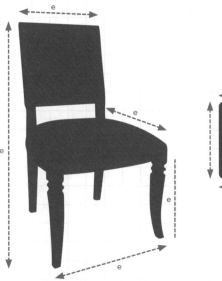

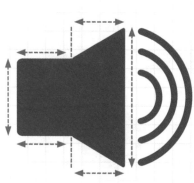

Putting the **E** in STEM Teaching and Learning

MORE EVERYDAY ENGINEERING

Putting the E in STEM Teaching and Learning

Richard H. Moyer and Susan A. Everett

Photography by Robert L. Simpson III

NSTApress

National Science Teachers Association

Arlington, Virginia

National Science Teachers Association

Claire Reinburg, Director
Wendy Rubin, Managing Editor
Rachel Ledbetter, Associate Editor
Donna Yudkin, Book Acquisitions Coordinator

ART AND DESIGN
Will Thomas Jr., Director

PRINTING AND PRODUCTION
Catherine Lorrain, Director

NATIONAL SCIENCE TEACHERS ASSOCIATION
David L. Evans, Executive Director
David Beacom, Publisher

1840 Wilson Blvd., Arlington, VA 22201
www.nsta.org/store
For customer service inquiries, please call 800-277-5300.

NSTA is committed to publishing material that promotes the best in inquiry-based science education. However, conditions of actual use may vary, and the safety procedures and practices described in this book are intended to serve only as a guide. Additional precautionary measures may be required. NSTA and the authors do not warrant or represent that the procedures and practices in this book meet any safety code or standard of federal, state, or local regulations. NSTA and the authors disclaim any liability for personal injury or damage to property arising out of or relating to the use of this book, including any of the recommendations, instructions, or materials contained therein.

LIBRARY OF CONGRESS CATALOGING-IN-PUBLICATION DATA
Names: Moyer, Richard. | Everett, Susan A.
Title: More everyday engineering : putting the E in STEM teaching and
   learning / by Richard H. Moyer and Susan A. Everett ; photography by
   Robert L. Simpson III.
Other titles: Everyday engineering
Description: Arlington, VA : National Science Teachers Association, [2016] |
   Includes bibliographical references and index.
Identifiers: LCCN 2016026297 (print) | LCCN 2016027811 (ebook) | ISBN
   9781681402789 (print) | ISBN 9781681402796 (e-book) | ISBN 9781681402796
   (pdf)
Subjects: LCSH: Inventions--Popular works. | Engineering--Popular works.
Classification: LCC T47 .M726 2016 (print) | LCC T47 (ebook) | DDC 620--dc23 LC record available at *https://lccn.loc.gov/2016026297*

The Next Generation Science Standards ("NGSS") were developed by twenty six states, in collaboration with the National Research Council, the National Science Teachers Association and the American Association for the Advancement of Science in a process managed by Achieve, Inc. For more information go to *www.nextgenscience.org*

CONTENTS

ABOUT THE AUTHORS

RICHARD H. MOYER is emeritus professor of science education and natural sciences at the University of Michigan–Dearborn, where he developed and taught courses in the science education program. He is a co-author of McGraw-Hill's *Inspire Science* and *Science: A Closer Look*. He and Dr. Everett are co-authors of a college science methods book as well as the original *Everyday Engineering*. Dr. Moyer has conducted hundreds of teacher workshops both nationally and internationally.

SUSAN A. EVERETT is a professor of science education and chair of the Department of Education in the College of Education, Health, and Human Services at the University of Michigan–Dearborn. Dr. Everett, with Dr. Moyer, is the co-author of the featured column "Everyday Engineering" in the National Science Teachers Association's middle-level journal, *Science Scope*. Dr. Everett regularly teaches elementary science methods courses, graduate-level research courses, and inquiry-based Earth science classes.

ACKNOWLEDGMENTS

WE OWE A great deal to Inez Liftig, the longtime editor of *Science Scope*. Her knowledge and expertise in middle-level curriculum and learners are unsurpassed in our view. She helped us shape our ideas related to curriculum integration, recognizing differences among the teaching styles of our readers, and supported our attempts to make connections from the historical to the present day. Inez has a strong environmental education focus and many times shared ideas with us to further our thinking about how what we engineer has an effect on the Earth. She encouraged our attendance at yearly *Science Scope* advisory board meetings, which provided us with valuable feedback from the leaders in middle-level science education across the country. Through this experience we met Patty McGinnis, then a National Science Teachers Association (NSTA) middle-level director, who invited us to participate in the *Meet Me in the Middle* conferences within the NSTA conference where we shared engineering ideas with hundreds of middle-level science teachers. There is no doubt that without Inez's influence, *Everyday Engineering* and now *More Everyday Engineering* would be less than they are.

We would also like to recognize our colleagues and students at the University of Michigan–Dearborn. During several different semesters, in addition to methods class micro teaching, preservice students taught our lessons at STEM nights at various metropolitan Detroit schools as part of their clinical experiences. Our colleagues in the Inquiry Institute at the university were always quick to inquire about our most recent engineering project with a "What are you making?" or "What are you taking apart now?" In particular, we want to single out chemist and science educator Charlotte Otto, who was invaluable in helping us interpret some complex chemistry (plastic polymers among others) at an understandable middle-level manner. We have always been able to count on our photographer and illustrator, Robert Simpson III. He has a unique talent that allows him to capture the interesting "engineering parts" of our materials through his photographs. Without his photography, we would not be able to communicate our work to teachers; his photography truly is worth more than the proverbial thousand words.

Our editors at *Science Scope* certainly make us better writers as well. Led by Ken Roberts, both Kate Lu and Janna Palliser ask us important questions to ensure that we are really saying what we mean—so that our lessons will (hopefully) always be clear to our readers. Even when we do not, Kate and Janna tend to the details, providing clarity and accuracy for the teachers and students using the investigations. Finally, Claire Reinburg, director of NSTA Press, and Wendy Rubin, managing editor, have once again taken several years' worth of distinct *Science Scope* columns and magically put them together into this seamless volume, *More Everyday Engineering*.

—Richard H. Moyer and Susan A. Everett

CHAPTER 1

INTRODUCTION

Engineering Is Everyday

FOUR YEARS AGO when we wrote the introduction to *Everyday Engineering: Putting the E in STEM Teaching and Learning* (Moyer and Everett 2012), we began with a discussion of what we mean by "everyday" engineering. We had spent time asking our colleagues what came to mind when they thought of engineering. This resulted in a wide range of responses, mostly technical. "Everyone recognizes that such things as computers, aircraft, and genetically engineered plants are examples of technology, but for most people, the understanding of technology goes no deeper" (ITEEA 2007, p. 22). Since the outset of this project, we have been fascinated by the engineering required to make the simplest things we use every day—from snap fasteners to string and rope to throwaway ear buds. It's easy to see that lasers, automobiles, and computers must be engineered. However, because of the commonness of many of the items we use each day, we may lose sight of the fact that those technologies had to be invented, designed, and tested by someone as well.

Engineering is the process we use to develop solutions to the problems humans face. It is usually an iterative or cyclical process, working to improve earlier designs. In the design process, engineers must consider the criteria to be met, as well as any constraints, to successfully solve the problem. For example, even ice cube trays must be designed. What are the criteria you might have for your ice cubes? Do you prefer larger or smaller ice cubes? Perhaps you like ice chips rather than cubes? Once the type of ice preferred is determined, if you were the engineer, you would need to consider the tray's design criteria and constraints. Larger cubes have less surface area than an equal mass of smaller ones and will thus be less likely to dilute your beverage—but one constraint is that they will take longer to cool the drink. The smaller cubes will cool your drink significantly faster but will water it down. Furthermore, the larger cubes will take longer to freeze in your refrigerator, whereas the smaller ones will freeze more quickly. You cannot have it both ways. Engineers must always find a balance between criteria and constraints. You may have noticed that most ice cube trays are of a similar, standardized size and shape. If you insist on a differently sized or shaped ice cube for *your* needs, then you must purchase a more specialized ice cube tray and be willing to accept its constraints.

Next Generation Science Standards

Inclusion of engineering and technology into the science curriculum by the *Next Generation Science Standards* (*NGSS;* NGSS Lead States 2013) has been widely noted, but it is not an especially new construct within the field of science education. Starting with the Sputnik era of the 1950s, which led to the plethora of "alphabet" revisionist science programs, to Project 2061 of the American Association for the Advancement of Science (AAAS 1993) in the 1980s and 1990s to the *National Science Education Standards* of the National Research Council (NRC 1996), there has been a steady increase in the amount of engineering and technology recommended for inclusion in science education. Only with the publication of the *NGSS*, however, has engineering

TABLE 1.1	MS-ETS1 Engineering Design

Students who demonstrate understanding can do the following:	
MS-ETS1-1.	Define the criteria and constraints of a design problem with sufficient precision to ensure a successful solution, taking into account relevant scientific principles and potential impacts on people and the natural environment that may limit possible solutions.
MS-ETS1-2.	Evaluate competing design solutions using a systematic process to determine how well they meet the criteria and constraints of the problem.
MS-ETS1-3.	Analyze data from tests to determine similarities and differences among several design solutions to identify the best characteristics of each that can be combined into a new solution to better meet the criteria for success.
MS-ETS1-4.	Develop a model to generate data for iterative testing and modification of a proposed object, tool, or process such that an optimal design can be achieved.

Source: NGSS Lead States 2013.

been cited as a fundamental part of *A Framework for K–12 Science Education* (*Framework*; NRC 2012). The *Framework* is composed of three dimensions: disciplinary core ideas, science and engineering practices, and crosscutting concepts. For the first time, engineering is included as both a core idea and as a set of practices. The eight science and engineering practices are intertwined with one another but differ in two basic ways—science focuses on questions, whereas engineering considers problems; science seeks explanations, whereas engineering works toward solutions. The practices are essential for students to understand how the processes of science and engineering take place. To this end, the authors of the *Framework* note, "We are convinced that engagement in the practices of engineering design is as much a part of learning science as engagement in the practices of science" (NRC 2012, p. 12).

This interconnection of science and engineering has significant history. Often, science is furthered by engineering developments. Conversely, engineering is sometimes advanced by scientific knowledge. For example, the development of the scanning, tunneling microscope aided scientists in learning more about the atom (nanoscience), which enabled engineers to develop technologies at a very small scale (nanotechnology). Ideally, with the acceptance of the *NGSS*, students will have greater opportunities to experience these connections of engineering and science.

In Table 1.1, we list the engineering performance expectations (MS-ETS1) for middle-level students. The performance expectations are written based on the core ideas, the practices, and the crosscutting concepts. Let's consider our ice cube tray design problem noted earlier, which meets two of the expectations in the box. Students are required to consider the criteria and constraints of their problem—do they want ice cubes that last a long time or ones that cool their beverage quickly, as suggested in MS-ETS1-1? Students then test three different ice cube tray designs, collecting data for analysis to determine how well each tray meets the criteria and constraints of the problem (MS-ETS1-2).

TABLE 1.2	Book topics organized by area

Engineering Area	Topic
Engineering Amusements	• Toying Around With Windups • Budding Sound Engineers: Listening to Speakers and Earbuds • An In-Depth Look at 3-D
Engineering Materials	• Producing Plastic ... From Milk? • If It's Engineered, Is It Wood? • UV or Not UV? That Is a Question for Your Sunglasses • Why the Statue of Liberty Is Green: Coatings, Corrosion, and Patina
Engineering at the Retail Store	• Should Ice Be Cubed? • It's Stuck on You • Queuing Theory—Is My Line Always the Slowest?
Engineering Ordinary Things	• Keeping It Together—Fascinating Fasteners • Twisting and Braiding—From Thread to Rope • Sitting Around Designing Chairs

Everyday Engineering Activities

Like the original *Everyday Engineering* book, this volume is based on our regular column, "Everyday Engineering," published in the National Science Teachers Association (NSTA) middle-level journal, *Science Scope*. The first activity, "What Makes a Bic Click?" appeared in spring 2009, and the first book followed in spring 2012. Each article in the series makes use of a 5E learning-cycle format to investigate an everyday engineering problem. From the outset, we have attempted to bring forth activities that were not commonly published elsewhere. That is, we have tried to go beyond bridge-building and egg-dropping activities, preferring instead to have students deal with problems that are part of their everyday lives—for example, Band Aids or sunglasses.

Each installment is an investigation composed of a complete lesson that includes in-depth teacher background information, expected sample data, a materials list, a student activity sheet for recording results, and a short historical background. Detailed yet easy-to-understand explanations are provided in the teacher background information, taking into account that some

users may have a more limited technical background in science or engineering. To this end, sample data are always provided so teachers have an idea of what to expect when students conduct the activity. Also, sample student activity sheets are provided for teachers whose students require additional scaffolding and structure. The activities make use of simple, inexpensive materials that can be found around the science classroom or the local supermarket or dollar store. The historical background provides a human element to engineering, something that may escape some students. We feel this background is important to help students realize that engineering exists to solve human problems, especially those students who do not have an interest in STEM disciplines. Furthermore, each investigation is supported by detailed color photography. The photography offers additional information for the reader—an inside view of an earbud or detail of the different ways to twist fibers together into threads and ropes.

The topics in *More Everyday Engineering* are organized into four sections: Engineering Amusements, Engineering Materials, Engineering at the Retail Store, and Engineering Ordinary Things (see Table 1.2, p. 3). Again, it is our desire that the topics will appeal to all students and help them understand that engineering truly is a part of their everyday lives.

The activities presented in *More Everyday Engineering* focus on three different aspects of engineering—designing and building, reverse engineering to learn how something works, and constructing and testing models. We point this out to emphasize that engineering education must include more than just the design process. The "Budding Sound Engineers: Listening to Speakers and Earbuds" activity is an example of the design and building process. After looking at an opened earbud, students must design and construct a simple speaker out of disposable plates and cups, magnetic wire, and magnets. In the activity, "Toying Around With Windups," students reverse-engineer a windup toy to see how it stores energy in a spring and transfers energy through a series of small gears to make the toy move. Finally, students make a model of plywood, a type of engineered wood, to test how the layering of the plies increases strength in the activity, "If It's Engineered, Is It Wood?"

STEM

While the standards (and our book title) emphasize the connections between science and engineering, the inclusion of the four STEM areas—science, technology, engineering, and mathematics—has also received considerable attention. Our intention in many of the lessons in *More Everyday Engineering* is to help teachers integrate the different disciplines when appropriate. For example, in "Toying Around With Windups," students are investigating the gearing mechanism in windup toys that transfers the energy stored in a spring into some sort of motion. Students take apart a windup toy to see how the gears transfer energy from the spring. They construct gears from foam core boards to learn the relationships between different sizes and types of gears. In this example of a STEM lesson, the science concepts investigated are energy transfer and motion. The technology is constructing gears (including

cluster gears), and the engineering problem requires the students to design a system of gears to turn another gear three times in the same direction as the first gear. Finally, the mathematics involves the ratios of teeth in adjacent gears and how these ratios are related to the speed of rotation of the interacting gears.

Conclusion

In the conclusion to *Everyday Engineering*, we noted that we hoped we inspired you to look at your garlic press in a different way than you had before. We still have that wish. Be curious about the things you come across daily—a stapler, a doorknob, or a pepper mill. How do they work? How were they made? How could you improve on their designs? We also hope *More Everyday Engineering* helps you pass this innovative spirit along to your students so they recognize and appreciate that engineering is truly *everyday*.

References

American Association for the Advancement of Science (AAAS). 1993. *Benchmarks for science literacy*. New York: Oxford University Press.

International Technology and Engineering Educators Association (ITEEA). 2007. *Standards for technological literacy: Content for the study of technology*. Reston, VA: ITEEA.

Moyer, R., and S. Everett. *Everyday engineering: Putting the E in STEM teaching and learning*. Arlington, VA: NSTA Press.

National Research Council (NRC). 1996. *National Science Education Standards*. Washington, DC: National Academies Press.

National Research Council (NRC). 2012. *A framework for K–12 science education: Practices, crosscutting concepts, and core ideas*. Washington, DC: National Academies Press.

NGSS Lead States. 2013. *Next Generation Science Standards: For states, by states*. Washington, DC: National Academies Press. *www.nextgenscience.org/next-generation-science-standards*

PART 1

Engineering Amusements

CHAPTER 2

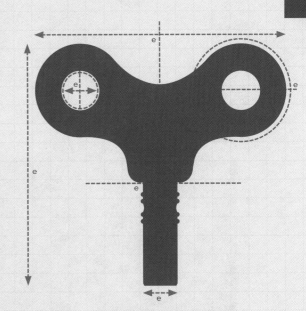

TOYING AROUND WITH WINDUPS

YOU WIND IT up, put it on the table, and away it walks. Or jumps. Windup toys have two things in common—a spring to store energy and gears to transmit it. However, a quick look at most windup toys does not reveal either the spring or the gears, which are almost always hidden inside, as are most of the gears you use every day. Analog clocks and watches use gears to turn the hands; many window treatments, such as mini-blinds or Roman shades, have gears in their mechanism; and automobile transmissions, blenders, and food processors all have internal gears. A few everyday objects have more readily visible gears— bicycles, old-style eggbeaters, corkscrews, and manual lawn mowers.

In this *More Everyday Engineering* activity, we will use windup toys to introduce students to how gears interact. Students will reverse engineer a windup toy to see how the gears transmit motion, thus causing the output motion of the toy. They will use their own gears of different sizes to discover how, when two gears are meshed, turning one affects the rotation of the other. The *Next Generation Science Standards* (*NGSS*) state, "When two objects interact, each one exerts a force on the other that can cause energy to be transferred to or from the object" (NGSS Lead States 2013; PS3.C). This lesson can also reinforce the concept of energy transfer. Kinetic energy is transferred to the device as it is wound and then stored as potential energy in

a spring. When the spring is released, the potential energy is transferred to kinetic again. Students may assume that the energy is gone once the motion stops; in reality, the energy is transferred to thermal energy and sound energy. Furthermore, the *NGSS* crosscutting concept Systems and System Models (MS-PS3-2) includes "Models can be used to represent systems and their interactions—such as inputs, processes, and outputs—and energy and matter flows within systems" (NGSS Lead States 2013). Working with gears in a simple windup toy offers the opportunity for students to analyze a system where energy is stored and transferred, resulting in some output motion.

Historical Information

References to gears can be traced back at least to Archimedes in the third century BCE. Early Egyptians used gears to lift water. The engineering problem they faced was that they needed to lift the water vertically while their oxen were walking in a horizontal plane. To accomplish this, they attached wooden pegs to two wheels mounted at right angles to one another (Torrey 1945). Throughout history, gears have been used to transfer motion in many different ways and for many purposes, including in grain mills and waterwheels. You may be surprised, however, to know that windup toys are also quite old. Early windups were not inexpensive

FIGURE 2.1	The inner workings of a toy clock

toys, but intricate mechanical devices. In Germany in the 1400s, Karel Grod made a number of flying windup toys, including a fly and an eagle that he would release at official gatherings. In 1509, Leonardo da Vinci created a windup lion (Wulffson 2000). By the 1600s, French artisans were making expensive windups out of silver, obviously intended for adults rather than child's play. Windup musical boxes were made by Swiss clockmakers in the early 1700s, again intended for adults (Victoria and Albert Museum 2012). It was about a century later that music boxes became toys for children. Metal windup toys began to appear in the United States in the late 1800s (Sobey and Sobey 2008). Today, of course, there is a plethora of inexpensive, imported plastic windups.

Investigating Gears (Teacher Background Information)

Materials

For the Engage activity, each group of three to four students will need a windup toy that they can eventually take apart (see Figure 2.1). Therefore, you want to obtain toys that are screwed together rather than glued. Windup toys are readily available online, as well as in dollar stores for about a dollar apiece. Although students should be able to reassemble the toys for reuse, you should have extras on hand for subsequent classes. Simpler toys are more apt to be put back together, as they have fewer parts. Thus, you might want to obtain those that have only one function (i.e., swim, walk, or roll). Each group will also need a small screwdriver (slotted or Phillips, depending on the device) suitable to disassemble the toys. If you need to purchase these, inexpensive ones are available for about a dollar (or less) at dollar stores and discount stores. You may also wish to provide a shoebox lid or other shallow container for student groups to use while disassembling the toys to keep track of all the small pieces.

For the Explore stage, each group will need six gears of three different sizes. The ones we are using are of a ratio of 24 to 18 to 12 teeth. You may already have gear kits in your lab. If not, you can make inexpensive ones from foam-core board as we have done here. Each group will require two sheets of foam-core board approximately $19.05 \times 25.4 \times 0.3175$ cm ($7.5 \times 10 \times 0.375$ in.). You can easily cut these ahead of time with a small utility knife. The gear patterns (see Figure 2.2; available for download at *www.nsta.org/more-engineering*) will be glued to one, and the other will be a base for attaching the gears. You will also want to glue the gear patterns to the foam-core board ahead of time so the glue has enough time to dry. Foam-core board is available at some dollar stores for about a dollar for a 76.2×50.8 cm (30×20 in.) sheet. Each sheet will accommodate four groups of students. Each group will need three or four paper clips to attach the gears, scissors, tape, and a glue stick.

FIGURE 2.2 Gear patterns

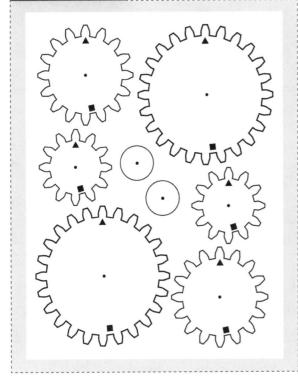

FIGURE 2.3 Multiple views of a windup toy

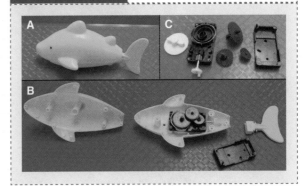

For the Extend stage, provide each group with a simple, everyday device that contains accessible gears. Such devices include eggbeaters, windup kitchen timers, clocks or old watches, mini-blind mechanisms, blenders, dry-line-type correctors, and windup toys.

Engage

Safety note: Students should wear safety glasses for this activity. To gain some understanding of students' prior knowledge, have them share and discuss windup toys and devices with which they are familiar. In the Engage stage, students will examine and take apart a windup toy to observe how it operates (see Figure 2.3). It is advantageous for students to delay taking apart the gearbox in the toy until they have more experience with the function of gears. Also, if it is possible, have students take a reference photo of the gear arrangement prior to disassembly (in the Extend stage) to aid in putting it back together.

You might want to wait to share the following information until the Explain stage of the lesson. Most windup toys operate essentially the same—energy is transferred to the toy by winding a stem, and then the energy is stored in a coiled spring. Like a music box, most windup toys start operating as soon as you let go of the stem. As the spring unwinds, it turns a shaft to which a gear is attached, which in turn meshes with one or more other gears. Often the gears change the direction or the speed of the motion in order for the toy to walk or jump, for example. The tail of the shark toy in Figure 2.3 will move back and forth, enabling the shark to "swim."

Rather than only using text, students will find it easier to record their thinking in more depth by making annotated drawings to explain how the various devices function. If the technology is available, you may also want students to use digital photography or videography to document the motion of their devices. Through discussion, encourage the class to pose questions about the interaction of the gears.

FIGURE 2.4 Gear setup

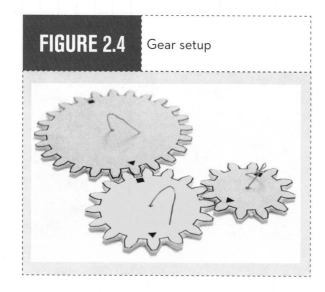

Explore

It is recommended that students first cut completely around each gear in a circle and then cut out the individual spaces between the teeth. To have a model ready for your students, prepare in advance a sample of gears (small, medium, and large) attached to a base of foam board with the teeth meshed (see Figure 2.4). It is important to bend the upper, projecting end of the paper clip down to apply a bit of pressure so the gears will stay flat on the board as they turn.

Students will use the gears to investigate several questions. They will first determine that adjacent gears always turn in opposite directions. In multiple linear gear arrangements, they will discover that every other gear turns in the same direction. They will also use the gears to determine the effects of different gear sizes. Sample data are provided in Table 2.1. Students are also asked to turn the smaller gear and note the effect on the medium and larger gears (see sample data in Table 2.2).

In the second part of the Explore stage, students will use a glue stick to combine gears (or spacers) on top of one another to investigate stacked gears, which are known as *cluster gears* (see Figure 2.5).

Explain

Students should discover that the number of times a gear turns depends on the number of its teeth and the number of teeth on the gear that is turning it. For example, when students turned the gear with 24 teeth

TABLE 2.1 Sample data: How many times do different-sized gears rotate when the large gear is turned?

Number of turns of large gear (24 teeth)	Number of turns of medium gear (18 teeth)	Number of turns of small gear (12 teeth)
1	1.5	2
2	3	4
3	4.5	6
12	18	24

TABLE 2.2 Sample data: How many times do different-sized gears rotate when the small gear is turned?

Number of turns of small gear (12 teeth)	Number of turns of medium gear (18 teeth)	Number of turns of large gear (24 teeth)
1	0.75	0.5
2	1.5	1
3	2.25	1.5
12	9	6

FIGURE 2.5 Cluster-gear setup

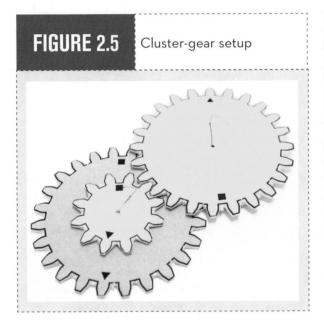

FIGURE 2.6 Windup toy with visible coil spring and gears

1 time, it caused the 12-tooth gear to turn 2 times. Note that the ratio of 24:12 reduces to 2. Likewise, when they turned the large, 24-tooth gear 3 times, the 18-tooth gear turned 4.5 times. Again, note that the ratio of 24:18 is 1.5, and thus 3 turns of the large gear yields 4.5 turns of the medium gear ($3 \times 1.5 = 4.5$).

Similarly, when students used the small gear as the driving gear to turn the other gears, they discovered that 1 turn of the small gear resulted in 0.75 turn of the medium and 0.5 turn of the large gear. Note here that the ratio of the gears' teeth is 12:18 for the small to the medium, which reduces to 0.75. The ratio of the small to the large is 12:24, which is 0.5.

Because the cluster gears are rigidly attached to one another and both turn in the same direction, students will discover that adjacent gears turn in opposite directions, as above. When they turn the large single gear one revolution, they will find that the other large gear will turn twice. This is because the single, large, 24-tooth gear is meshed with the smaller, 12-tooth gear that is rigidly connected to the larger bottom gear. Thus they both turn two times for every time the larger single gear turns. The reason for combining gears in this way is to change the speed that the driven gears turn. Note how this is different from one large gear driving another large gear—one revolution of the large gear would then result in the second large gear turning only once, as well.

Extend

For the Extend stage, you can reuse the windup toys from the Engage stage. You might also add some other mechanisms that have a spring and a gear, as noted in the "Materials" section (p. 8). At this time, you can have the groups carefully take apart the spring/gear assemblies to determine how the system transfers the input energy to the output motion. In Figure 2.6, you can easily see the coil spring that will contract as it is wound to store energy. The energy is released as the spring unwinds and causes the shafts with the gears to turn. Note also that the gearing results in the increased speed of the axles, because larger gears are turning smaller ones.

Have students analyze the input and output for their devices. In most cases, the input will be the turning of a crank that winds a (usually coil) spring. In this way, the input kinetic energy is transferred to potential energy that is stored in the spring. In our example of the toy shark (Figure 2.3), note that the gearing increases the speed of the tail's output motion. It also changes the plane of the motion—the input winding is in the horizontal plane, whereas the tail motion is in the vertical. In Figure 2.3C, you can observe an example of a plastic cluster gear; in this case, they are molded together rather than glued.

Evaluate

For the Evaluate stage, students will consider a slightly different set of three gears. One way to design a system to increase the speed by a factor of 3 and still have the output in the same rotational direction would be to use the large, 18-tooth gear to drive the 12-tooth gear, which in turn drives the 6-tooth gear. Because there are three gears in a row, the third one will rotate in the same direction as the first. The ratio of the driving gear to the small gear is 18:6, or 3. If you remove the middle gear, then the ratio remains the same, but the smaller gear will now turn in the opposite rotational direction.

Conclusion

This activity incorporates all four STEM areas. The science content covers energy storage and transfer.

The technology is made up of the windup toys and other devices with gears, and the engineering is how gears are designed to interact in a system. The ratio of the number of teeth in the gears constitutes the mathematics. While most of the gears we use may not be readily visible, they play a vital role in our everyday lives, from a simple windup toy to a complex automobile transmission.

References

NGSS Lead States. 2013. *Next Generation Science Standards: For states, by states.* Washington, DC: National Academies Press. *www.nextgenscience.org/next-generation-science-standards*

Sobey, E., and W. Sobey. 2008. *The way toys work: The science behind the Magic 8 Ball, Etch A Sketch, boomerang, and more.* Chicago: Chicago Review Press.

Torrey, V. 1945. Wheels that can't slip. *Popular Science* 146 (2): 120–125.

Victoria and Albert Museum. 2012. Moving toys. *http://media.vam.ac.uk/media/documents/legacy_documents/file_upload/17576_file.pdf*

Wulffson, D. 2000. *Toys! Amazing stories behind some great inventions.* New York: Henry Holt.

ACTIVITY WORKSHEET 2.1 — Get in Gear!

Engage

1. Observe the windup toy your teacher has provided, noting its movement. List as many other windup toys or devices as you can.

2. How do you think the windup toy operates? Record your ideas by making a drawing of what you think is inside the toy.

3. Put on your safety glasses. Use the screwdriver provided to take the toy apart, being careful not to break or lose any of the pieces. Again, turn the crank and observe the motion of the mechanism. What do you notice about the output motion compared to the turning of the crank, which is the input motion?

4. How do the workings of the toy compare with your initial thinking? Modify your drawing if necessary. You may have noticed small gears and possibly a metal spring in the workings. In the activity that follows, you will explore how gears change motion.

Explore

1. You will first need to prepare some gears for use in your exploration. Your teacher will provide a foam-core board with different-sized gears already drawn on it. Cut out each shape, being careful to cut around the teeth of the gears. Open one end of a paper clip and use it to poke a hole through the center of each shape.

2. Select three different-sized gears. Place one of the gears on the plain piece of foam-core board and use the paper clip to make a hole through the center of the gear and all the way through the board, as well. Remove the paper clip and, from the bottom of the board, insert the open end through both holes. Tape the clip in place on the bottom of the board. Finally, bend the open end of the clip down to hold the gear in place. Make sure the gear will turn easily.

3. Repeat this process with the other two gears, making sure the teeth mesh together so that when you turn one gear, it causes the one next to it to turn as well.

4. Use your gears to investigate the following questions:

    a. If you turn one gear clockwise, what direction do the other two turn? What if you turn the first gear counterclockwise?

    b. Turn the large gear one complete revolution. Count the number of times each of the other two gears turn. Record your data in a table like the one on the next page.

    c. Complete another table, but turn the small gear and count the revolutions of the other two.

5. Now investigate what happens when gears are assembled with one attached to the top of another. With a glue stick, attach a small gear on top of the largest gear and mount it on the foam-core board with a paper clip. Glue a circular spacer to the bottom of a second large gear and mount it on the board so that its teeth mesh with the teeth of the adjacent smaller gear. You have created one double gear, known as a cluster gear, and a second single gear with a spacer on the bottom so that it can mesh with the top gear of the cluster gear.

6. Use this gear assembly to answer the following questions:

    a. If you turn the cluster gear clockwise, what direction does the single large gear turn? What if you turn it counterclockwise?

b. Turn the single large gear one complete revolution. How many times does the large gear of the cluster gear turn?

c. How might other combinations of gears work? Test at least one other combination to find out.

Explain

1. In general, how does the direction a gear turns affect the direction of the gear it is turning? What if there are more than two gears meshed together?

2. What is the relationship between the number of teeth and the number of times a meshed gear turns? Use your data tables for the large and small gears to help answer this question.

3. In the second part of your exploration, what was the relationship between the number of times the two elevated gears turned? What about the number of times the two large gears turned? Why might you want to put gears together to form this type of system?

Extend

1. A system is a collection of parts that interact with one another to accomplish some task or other outcome. Your teacher will provide you with a device that contains gears. Determine how the mechanism works as a system, making close observation of the gears. Draw a sketch to record your thinking. What is the outcome of your system of parts?

2. What do you have to do to start each device? What is the resulting output motion of the device? How is energy transferred?

3. Based on your sketch, what happens between the input and output motions? How many gears do you see? Describe the direction they are turning. Locate any cluster gears that are stacked together.

4. Present your findings to the class.

Evaluate

How would you design a system of gears that ended with a motion that is in the same rotational direction as the input motion but is turning three times faster? Imagine you had three gears with which to do this—one with 18 teeth, one with 12 teeth, and one with 6 teeth. Make a sketch of your design and then explain. How would you need to change your design if you wanted the motion to be in the opposite rotational direction of the first gear?

Number of turns of large gear (24 teeth)	Number of turns of medium gear (18 teeth)	Number of turns of small gear (12 teeth)
1		
2		
3		
12		

CHAPTER **3**

BUDDING SOUND ENGINEERS
Listening to Speakers and Earbuds

IT SOMETIMES SEEMS as if many of us are plugged in much of the time—to earbuds, that is. Part of this earbud proliferation is undoubtedly due to their low cost, making them nearly disposable (see Figure 3.1). Although higher-end earbuds exist, inexpensive ones can be purchased for less than a dollar. They are often given away on planes and tourist buses. Have you ever wondered how they work? How are earbuds functionally different in design from any other headphones? Are headphones fundamentally different from loudspeakers?

In this 5E learning-cycle lesson, students construct a simple speaker consisting of a coil of wire, a magnet, and a disposable cup for a diaphragm that is connected to an MP3 player. Students then "[e]valuate competing design solutions" (MS-ETS1-2; NGSS Lead States 2013) to determine how to improve the performance of their speakers. The related science content requires students to recognize that sound vibrations are transmitted from one medium to another: "A sound wave needs a medium through which it is transmitted" (MS-PS4-2; NGSS Lead States 2013). The magnet and the coil convert the electrical current from the MP3 player into a constantly changing force against the bottom of the cup, matching the pattern of the originally recorded sound or music. This force causes the cup to vibrate in the same pattern and then transmit

| FIGURE 3.1 | Typical earbuds |

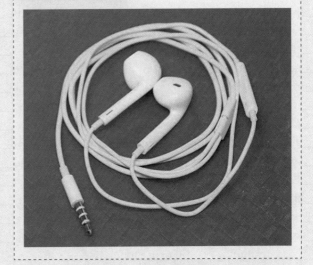

those vibrations into the air, which we perceive as a reproduction of the recorded sounds.

Historical Information

Some people believe that earbuds came on the scene with the introduction of the iPod in fall 2001. However, they were invented in 1891 by a French engineer,

Ernest Mercadier, who obtained a patent for in-ear headphones for use by telephone operators (Stamp 2013). American Nathaniel Baldwin further developed headphone technology at the onset of World War I, improvements that were used by military radio operators. The next major breakthrough in headphone development occurred in 1957, when stereophonic recordings were introduced. Noise-canceling headphones were initially developed at about the same time. In a precursor to what is happening today, after the 1979 introduction of the portable cassette player known as the Sony Walkman, people were walking around wearing earphones as they listened to their favorite music (Stamp 2013). (Earphones and earbuds essentially differ only in that earbuds fit into the ears, whereas earphones are worn over the ears.)

Today, the science of both loudspeakers and earphones is essentially the same, but the earliest loudspeakers were little more than a megaphone. Early phonographic horns did not use an electric current and did little to magnify sounds. They were used in recording devices from the 1880s to 1910, when the invention of the vacuum tube made the amplification of sound possible. Although the idea for loudspeakers was first proposed in the 1870s, the first modern loudspeaker was developed by C. W. Rice and E. W. Kellogg in 1921. They combined early technology into a workable design, and eventually their speakers were widely sold by the Radio Corporation of America, better known as RCA (Edison Tech Center 2012). This basic technology is still in use today.

Investigating Speakers and Earbuds (Teacher Background Information)

Materials

Have students work in groups of four for this investigation. If possible, during the speaker construction, have students work in pairs. Thus, for the Engage phase, you will need 12 pairs of earbuds for a class of 24. We recommend that teachers prepare the earbuds ahead

FIGURE 3.2 Inside of an earbud

of time: Wearing goggles, cut the earbuds from the wire and use pliers to gently open them, revealing the inside components—a wire coil, a magnet, and a plastic diaphragm (see Figure 3.2). Provide hand lenses so students can more easily observe the components of the opened earbud. Strip 1 or 2 cm from the cut end of each wire and retain the plug end for the Explore phase.

Students must wear safety goggles during the Explore phase of the lesson. For each group of four students, provide one MP3 player (two would be ideal so pairs do not have to share), two cables with earphone plugs that were cut from earbuds, two foam and two plastic cups of the same size (we used 16 oz. cups [about 480 ml]), four wires with alligator clips, four 6 m pieces of 30-gauge enameled (magnet) wire, and four doughnut magnets. Each group will also need some general supplies—an empty toilet-paper tube, tape, and a small (10 cm^2) piece of sandpaper. If you want each group to make its own coils or to experiment with the coils, you will need about 600 m of wire. Wire is available online or from electronic hobby stores for about $2 per 100 m. This wire is insulated with an enamel coating. Do not make the coils out of uninsulated wire, because each loop will short against the next. The sandpaper is used to remove the insulation from the ends of the wire to make a connection. The magnets should be just smaller than the diameter of the toilet-paper tube so they fit inside the coils on the bottom of the cups (see Figures 3.3 and 3.4). If insufficient MP3 players are available, set up several listening areas where students can test their speakers.

For the Extend phase of the lesson, you will need photos of various speakers, which can be obtained online by searching for loudspeaker images or something similar. If available, bring a speaker to class for students to examine. Such speakers may be found in outdated radios or music players often available at garage and yard sales. All materials except for the cups and tape could be reused from one class to another. Students will need to complete the Explore phase in one class session so that the materials may be reused.

Engage

Although students may use earbuds frequently, they probably have little knowledge of what is inside or how the internal components work together to reproduce sounds. In fact, students may not recognize any of the three major components—the coil, the magnet, or the diaphragm. It is important that students become aware of these parts prior to beginning the construction of their speakers in the Explore phase of the lesson.

Explore

Safety note: Students must wear safety goggles for the duration of the exploration.

The steps here are quite directed, leading students to build a simple speaker that can reproduce the weak signal of an MP3 player. Although the output from the MP3 is quite clear and distinct, it is not very powerful. For this reason, you may wish to have a couple of quiet "testing minutes" for students to listen to their cup

FIGURE 3.3 MP3 setup with cup, coil, and magnets

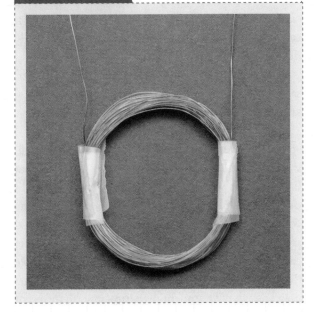

FIGURE 3.4 Wire coil

CHAPTER 3

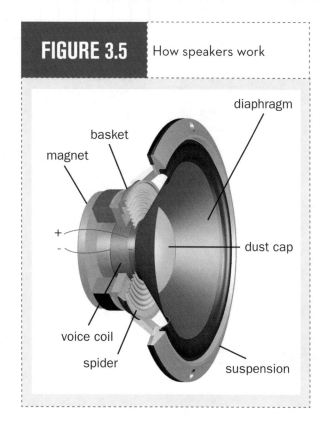

FIGURE 3.5 How speakers work

diaphragm materials, such as paper, plastic, and foam plates and bowls. For example, a metal coffee can works rather well, but a glass beaker does not.

Explain

Students will probably find that the plastic cups produce louder sound than the foam cups. We also found that the sound produced by a metal coffee can is louder than that from the lower-density foam cups. The higher density of the plastic and metal materials allows sound energy to be more easily transmitted from one particle to the next.

Discuss with students their ideas regarding how musical instruments produce sound. Though many students understand that musical instruments produce vibrations, they may not be aware of the transfer of sound energy that is entailed. For a guitar, for example, the string vibrates as it is plucked and then transfers this vibration to the wooden body of the guitar, especially the top plate, which then causes the air to vibrate in a similar manner, transmitting the sound to the listeners' ears.

A speaker produces sound in a somewhat analogous manner, although students will probably not be aware of something vibrating (like the guitar string) in the cup speaker. The interaction of the magnetic fields produced by the coil and the permanent magnet causes the bottom of the cup to vibrate in sync with the originally recorded sound. These vibrations are then transmitted (as are the vibrations of the guitar string to the guitar body) to the cup and then to the air. Students must infer that the cups are vibrating even though they cannot see or feel the vibration because, otherwise, they would not be perceiving the sound. You may wish to view (and possibly share with your students) the video "How speakers work" (see Resource, p. 19).

Extend

Using either photos or speakers, have students examine several types of speakers. Although some differences will exist, students should readily discover that the three major components are always present—a magnet, a coil, and a diaphragm (see Figure 3.5). Some, for example, may have

speakers. If students are having trouble getting their cup speakers to work, have them make sure that the alligator clips are properly attached and not touching each other (which would create a short), that the coil is securely attached, and that the two magnets are firmly held to the bottom of the cup.

After the successful construction of one speaker, we suggest changing the cup material (the diaphragm) and testing again. At this point, you may wish to have students design their own modifications to this basic speaker model—such as changing the cup material, the coil, or the magnet. You will probably find that little changes when using ordinary magnets or slight differences as you alter the number of turns in the coils (do not use less wire, however, as the lessened resistance could damage the MP3 player). Changing the cup material, however, does result in noticeable differences in the volume of the sounds produced. You may wish to have students experiment with different the coil inside the magnet. The material of the speaker cone or diaphragm may vary as well. The diaphragm is always, however, the part of the speaker that vibrates and transmits the reproduced sound to the air.

If you wish to continue this lesson in additional ways, ask students to reflect on the importance of speakers to society. Undoubtedly, they will be able to identify benefits of speakers, but you might want to challenge them to note some disadvantages, if possible, as well. You might also want to have students research modern developments in speakers. In addition, students may wish to investigate how speaker technology is used in other designed products such as hearing aids.

Evaluate

This activity allows you to assess whether students understand that the three major speaker components— coil, magnet, and diaphragm—interact to reproduce sound. It also provides an opportunity for students to participate in the creative aspects of engineering design. Students may suggest using materials such as pieces of paper, paper or plastic plates, file folders, or report covers. You may also wish to have students construct some of the designs they have drawn.

Conclusion

Through this investigation, we once again reinforce our guiding premise that engineering is often about the very simple things that people use every day. With the exception of the introduction of stereo and noise-canceling technology and improvements to overall audio quality, the basic function of the coil, magnet, and diaphragm has changed little since Rice and Kellogg produced the first speaker in 1921.

References

Edison Tech Center. 2012. History and types of loudspeakers. *http://edisontechcenter.org/speakers.html*

NGSS Lead States. 2013. *Next Generation Science Standards: For states, by states.* Washington, DC: National Academies Press. *www.nextgenscience.org/next-generation-science-standards*

Stamp, J. 2013. A partial history of headphones. *Smithsonian.com.* March 19. *www.smithsonianmag.com/arts-culture/a-partial-history-of-headphones-4693742*

Resource

How speakers work (YouTube video). *www.youtube.com/watch?v=DMxn3CPLe-A*

Engage

1. Do you wear earbuds when you listen to music? How do you think they work? What do you think might be inside them that allows you to hear sounds? Record your ideas.
2. Carefully examine the opened earbud that your teacher has provided. Do not attempt to take it apart any further. Make a sketch of what you see. Label any parts you think might help the earbud make sound.
3. Share your ideas with the class.
4. In this exploration, you will build and test different designs of a speaker that you can use with an MP3 player.

Explore

Safety note: You must wear safety goggles for the duration of this exploration.

1. The materials (wire for a coil, a magnet, and a plastic cup to transmit the sound) provided by your teacher can be used to construct a basic sound speaker. First, you will build a simple speaker, and then you will modify the design in an attempt to improve its operation.
2. Measure the diameter of the bottom of your foam cup. You must make a coil of wire that is just slightly smaller than this diameter so that the coil will fit inside the outer ridge on the bottom of the cup. To do this, you will need to use something similar to an empty paper-towel tube. Leave about 20 cm on each end of the wire and then tape the coil so it will not unwind. Using sandpaper, remove the enameled coating from the last 2 cm of each part of the wire that extends from the coil. Finally, use tape to attach the coil securely to the bottom of the cup.
3. Your teacher has also provided you with a wire cable from which the earbuds have been removed. Using an alligator clip, attach one end of your

coil to one of the bare wires extending from the earbud cable. Repeat with the other wire from your coil. Plug your cup speaker into an MP3 player. Switch the MP3 player on and turn up the volume. Hold a magnet to the bottom of the cup in the center of the coil. Pick up the cup and hold it near your ear. What do you hear?
4. Now, select a plastic cup and repeat the above procedure. Compare the sound produced by your two speakers. Record your findings. Your teacher may wish for you to change other aspects of the speaker design as well.

Explain

1. Share your results with the class. How well did each speaker produce the sound from the MP3 player?
2. Think about musical instruments with which you are familiar. How do they produce sound vibrations? For example, what part of a guitar vibrates to produce sound?
3. Now think about the cup speaker. What must be vibrating for it to make sound? How do you know?
4. What must the sound travel through so that you can hear it?

Extend

1. Look at other examples or images of speakers and earphones your teacher has provided.
2. How are they like the speaker you just made? What parts are similar?
3. What vibrates on these speakers to make sound?

Evaluate

Imagine you wanted to see whether a bigger speaker would produce louder sound from your MP3 player. What materials might you use to construct your speaker? Make a drawing of your design and label all the important parts. Write a short description of why you think your idea will work.

CHAPTER 4

AN IN-DEPTH LOOK AT 3-D

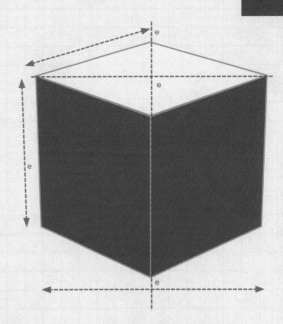

THREE-DIMENSIONAL (3-D) images—once limited to special movies and View-Master stereoscopes—are now commonplace. Many theaters show movies in 3-D, and video games, DVDs, and televisions all offer 3-D as well. Providing a perception of depth adds a sense of realism for the viewer. How, exactly, do we perceive three dimensions in real life, and how is that different from "seeing" three dimensions when looking at a flat screen?

There are several important visual clues that help us to perceive depth. One example is our experience with the relationship between size and distance. A second has to do with the fact that foreground objects may block some part of objects in the background, which is known as *obscuration*—a tree growing in front of a house blocks part of the house. Nonetheless, humans perceive three dimensions primarily because of binocular vision (Howard and Rogers 1996). *Binocular vision* means that each of two separate eyes sees a somewhat different image that is fused into one by the brain (Martin 2009). Binocular vision may lead to *stereopsis*, where the different images (each one shifted slightly to the right or left) seen by each eye are interpreted by the brain as depth. You can easily see how this works by placing a dot on a piece of paper and placing it within reach. Then, with one eye closed, try to quickly bring your finger from eye level straight

down on top of the dot. You likely will miss the first time you try this. Why? Hold your finger at arm's length and view with one eye. Now view with the other eye and note that, with respect to the background, the finger seems to move right or left. This is because of your binocular vision—each eye sees your finger from a slightly different horizontal position. Your brain interprets these two images and is able to perceive the third dimension, depth.

The engineering challenge for creating 3-D images is to devise systems in two dimensions (movie screens, televisions, etc.) that produce images that seem to be 3-D. To do this, a slightly different image must be shown to each eye, mimicking what occurs in stereopsis when we view the 3-D world. In this activity, students will investigate several different designs that enable the viewer to perceive 3-D images. The *Next Generation Science Standards* (*NGSS*) state that middle-level students should understand "[t]he iterative process of testing the most promising solutions and modifying what is proposed on the basis of the test results leads to greater refinement and ultimately to an optimal solution" (MS-ETS1-4; NGSS Lead States 2013). Methods for producing 3-D images have evolved using several techniques since their beginning in the 1840s. The science content of this lesson avails itself to inclusion in a unit on information processing

(LS1.D): "Each sense receptor responds to different inputs (electromagnetic, mechanical, chemical), transmitting them as signals that travel along nerve cells to the brain" (NGSS Lead States 2013).

Historical Information

Three-dimensional images have been around longer than you may think. The world's first photograph, actually known as a *heliograph* (literally, *sun writing*) was taken in 1826 by Joseph Nicéphore Niépce (known as Nicéphore Niépce) in France using an hours-long exposure time (*National Geographic*, n.d.). By 1838, Charles Wheatstone had developed the idea for stereo photography. He constructed the first device that allowed each eye to see a slightly different view of the same image. His idea, which used mirrors and prisms, was later simplified, and by the turn of the century, most homes had a stereoscope and a collection of cards for viewing (see Figure 4.1).

Motion pictures were developed in the later part of the 19th century. Thomas Edison perfected a viewer called a *kinetoscope*, a machine used to view a moving strip of still photographs that gave the illusion of motion (National Park Service 2012). Again, as with still photography, 3-D movies followed soon thereafter. In 1894, William Friese-Greene patented a system for viewing 3-D movies using two projectors side by side: "The first projected 3-D movies in America were screened on June 10, 1915, at the Astor Theater in New York City" (Zone 2005, p. ix). These first 3-D movies used a two-color system, called an *anaglyph*, where one image was blue and the other red. Theatergoers wore glasses with one blue and one red lens so that each eye saw a slightly different image. There was a boom in 3-D movies in the 1950s and again recently. While the technology has changed so that some systems rely on Polaroid lenses and others on synchronizing lenses, the basic idea has remained the same—each method presents slightly different images to each of the viewer's eyes.

FIGURE 4.1 Stereoscope

The View-Master was originally developed in 1938 when photographer William Gruber, who was taking 3-D photos of the Oregon Caves National Monument and Preserve, met Harold Graves, president of a photographic company. "Together, they would produce the View-Master, a new way of viewing tourist attractions in America" (Townsend 2011, p. 1). They introduced their new idea the next year at the New York World's Fair.

Investigating 3-D (Teacher Background Information)

Materials

For the Engage phase, each individual student will need a piece of scratch paper, a pencil or pen to make a dot, and a half sheet of standard writing paper (about 14 cm long; when rolled up, about 4 cm in diameter). Each student group will need some cellophane tape. For the Explore phase, each pair of students will need copies of the stereo photos shown in Figure 4.2 (image files available at *www.nsta.org/more-engineering*). Stereo

FIGURE 4.2 Stereo photos of a plant

FIGURE 4.3 Red/blue drawing

photos are two photos of the same object taken from a few centimeters apart horizontally. Thus, they replicate the view of the object that each of our eyes sees, given that our eyes are also horizontally separated by a few centimeters. (Students can also find stereo photos online by searching for "stereoscopic images," but searches should be monitored by the teacher.) Each pair of students will need some standard school supplies to build a 3-D viewing device—tape, scissors, paper, card stock, and so forth. In the Extend phase, each pair of students will need paper, red- and blue-colored pencils or markers, red/blue 3-D glasses, and a copy of a red/blue drawing such as the one shown in Figure 4.3 (image files available at *www.nsta.org/more-engineering*). Red/blue 3-D glasses are readily available online for about 15 cents apiece for cardboard and about $1.50 for plastic ones that could be sanitized with an alcohol wipe and reused (students should be sure that the alcohol has dried before putting on the glasses). The glasses are easy to locate online using the keyword search "red/blue 3-D glasses."

Finally, in the Evaluate phase, each group of four students will need a View-Master reel. The entire class can share one View-Master, although it would be best for each group to have its own. The reels are available for as little as $1 apiece, and View-Masters are about $5.

Engage

Before students are introduced to 3-D viewing, it is important for them to understand the functioning of binocular vision. Thus, we suggest that you initiate a discussion probing students' thinking about why some animals have two eyes (binocular vision). Essentially, there are three reasons for why we have evolved this way. One is simply to have a spare. Another is to broaden the field of vision: Animals with two frontal-facing eyes, like primates, can see a field of about 180°. Loss of one eye reduces this field by about 25%. Animals with eyes on the side of their head (lateral-facing orbits), such as zebras, have a much wider field of view—up to nearly 360° in some instances. Note that many prey animals have this wider view, and predators such as lions that hunt zebras have a more forward-looking field. This leads us to the third purpose of binocular vision of frontal eyes, which is to provide depth perception through stereopsis.

As noted above, students will have difficulty locating the position of a dot in front of them if they only use one eye to view it. In the second activity, students should find that the image of their finger seems to jump horizontally when they look at it with each individual eye. Again, this is because each eye has a slightly different view of the finger because the eyes are a few centimeters apart.

The final Engage phase activity reinforces the idea that each eye sees a separate image, and the brain fuses the two into a single image (Bhola 2006). In this case, one eye sees a hole, and the other sees the palm of the hand—when they are put together, there is the perception that there is a hole in the palm.

Safety note: If students experience dizziness or headaches from eyestrain at any time during the activity, have them alert you and refrain from continuing the activities. If you have any students who are visually impaired, consider how you may help them participate in the activities.

Explore

Distribute stereo photos to students (we have provided one in Figure 4.2, p. 23). If you simply put your hand between your eyes and bring the photos up to your hand, you should be able to perceive that the two separate images merge into one 3-D image. Given that the image is so close, it will likely be a bit out of focus. The challenge for students is to design a method so that each of their eyes will see only one of the images. Have students brainstorm several of their ideas and then share with the entire class. Each pair should select one idea to construct and test. Essentially, their engineering problem is to design a stereoscope. One simple method uses paper tubes with the ends folded over for safety (Figure 4.4). For reasons of sanitation, students should be encouraged not to share the paper tubes with others. Encourage students to go through an iterative process to improve their designs based on testing.

If you want students to take their own stereo photos, they can take two photographs of the same object from a distance of about 130 cm. Take the first photograph and then move the camera horizontally about 10 cm for the second.

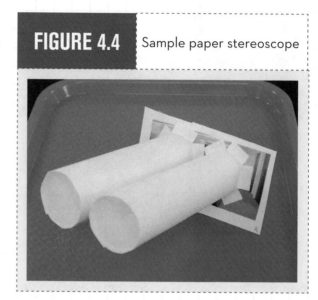

FIGURE 4.4 Sample paper stereoscope

Explain

Discuss the various products students designed to view the stereo photos. Students should recognize that to see the photos in 3-D, each of their eyes must see only one of the photos. To accomplish this separation, some sort of a barrier is required. Many designs may look a bit like a pair of binoculars and others like stereoscopes of old (Figure 4.1, p. 22). While your hand between your eyes will result in the perception of three dimensions, longer designs will allow the image to be seen more clearly. Students should now understand that a major result of binocular vision is that it is one way in which we can perceive depth.

To address the information-processing standard noted earlier, explain how the two images formed on the retina are transmitted via the optic nerves to the brain, where they are fused into one image. Nearer objects will be focused more to the outside of the retina and farther objects more toward the center. This displacement on the retina is perceived by the brain as distance. For further explanation and a retinal diagram of this concept, see Heesy (2009). You may also wish to invite an eye doctor to discuss depth perception and how the eye works with your class.

FIGURE 4.5   View-Master and reel

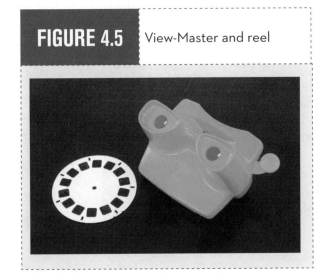

Extend

In contrast to the stereoscope students build, in this phase, they will use the red/blue glasses to perceive objects in three dimensions. Early 3-D movies made use of this technique. Again, the perception of depth is caused by each eye seeing a slightly different image. One image is drawn in red and the other slightly offset in blue (see Figure 4.3, p. 23). The eye with the blue lens will see only the red line, and the eye with the red lens will see only the blue, thus assuring that each eye sees a different image. In other words, the red line is "invisible" when viewed through only the red lens, and likewise, the blue line is "invisible" when viewed through only the blue lens. These two views are again fused by the brain into one 3-D image. NASA maintains a website, called Mars 3D, of anaglyphs of Mars (*http://mars.jpl.nasa.gov/mars3d*).

After students view the red/blue image, they should try to draw some of their own. Students may need to try several different shades of pencils (especially blue) to find the right one to match the lens in their red/blue glasses. Students will also need to experiment with how far apart (perhaps 1 or 2 mm) to make the red and blue lines in their drawings.

Evaluate

Have students look at images through a View-Master (Figure 4.5) and note that they are able to perceive three dimensions from a two-dimensional image. Have students examine the View-Master and corresponding reels of images. They should be able to determine again that a separate image is being shown to each eye. If they look through the View-Master with one eye at a time, they should notice (depending on the photo) that there are slight differences in each image. If they examine the images in the reels (there are two on opposite sides of the reel), they will see that these images are also slightly shifted.

Conclusion

Most of us probably do not think of entertainment and engineering in the same context. However, everything outside of the social and natural worlds was designed by someone. Design and innovation are critical to the development of technological devices, so even when watching a 3-D movie, engineering plays a role in our everyday lives.

References

Bhola. R. 2006. *Binocular vision.* Iowa City, IA: University of Iowa Healthcare. *http://webeye.ophth.uiowa.edu/ eyeforum/tutorials/Bhola-BinocularVision.htm*

Heesy, C. P. 2009. Seeing in stereo: The ecology and evolution of primate binocular vision and stereopsis. *Evolutionary Anthropology* 18 (1): 21–35.

Howard, I. P., and B. J. Rogers. 1996. *Binocular vision and stereopsis.* Oxford, UK: Oxford University Press.

Martin, G. R. 2009. What is binocular vision for? A bird's eye view. *Journal of Vision* 9 (11): 1–19.

National Geographic. n.d. Milestones in photography. *http://photography.nationalgeographic.com/ photography/photos/milestones-photography*

National Park Service. 2012. Motion pictures. Thomas Edison National Historical Park [National Park

Service website]. *www.nps.gov/edis/forkids/motion-pictures.htm*

NGSS Lead States. 2013. *Next Generation Science Standards: For states, by states.* Washington, DC: National Academies Press. *www.nextgenscience.org/next-generation-science-standards.*

Townsend, A. 2011. All-time 100 greatest toys: View-Master. *Time.* February 16. *http://content.time.com/time/specials/packages/article/0,28804,2049243_2048649_2049008,00.html*

Zone, R. 2005. *3-D filmmakers: Conversations with creators of stereoscopic motion pictures.* Lanham, MD: Scarecrow.

Resource

Cole, K. C. 1978. *Vision: In the eye of the beholder.* San Francisco: Exploratorium.

ACTIVITY WORKSHEET 4.1   Investigating 3-D

Safety note: Please be careful as you are completing the activities so that you do not poke yourself or anyone else in the eye. If you become dizzy during the activities, please alert your teacher.

Engage

1. Have you ever wondered why animals have two eyes? Share your ideas with the class. Try the activities below that focus on your vision and perception.

2. Make a dot on a piece of paper and put it on a table in front of you. Close one eye, hold your finger over your head, and quickly try to drop your finger down on top of the dot. Where did your finger land? Repeat with both eyes open. Which way were you more successful?

3. Look at your finger at arm's length with one eye. Then, without moving your hand, look at your finger with just the other eye. Describe what you see. Each of your eyes views the world from a slightly different position, a few centimeters apart. How does this help explain your observations?

4. Roll up a half sheet of paper into a cylinder about 14 cm long and about 4 cm in diameter. What do you predict you will see if you hold the cylinder to your eye and put your other hand next to the end of the cylinder so that one eye is looking through the cylinder and the other is looking at the palm of your hand? Why do you think this is what you will see? Try it. What do you see?

5. Based on these activities, have your ideas changed about why animals have two eyes? Give reasons for the changes in your idea. Discuss with the class.

Explore

1. Your teacher is going to give you two photos of the same object taken from slightly different positions—as if your eyes are in slightly different positions. Your challenge is to view the two photos so that you see just one 3-D image.

2. Discuss with your group how you can view the photos so that each of your eyes sees only one of the photos. Record your ideas and then share them with the entire class.

3. Test some of the ideas to see how well they work. Are you able to perceive the object in the photographs in three dimensions? How did you ensure that each eye saw only one image?

Explain

1. Consider the different methods you tested for viewing a 3-D image and share your reflections with the class.

2. Do any methods work better than others? Does each method work equally well for all people? Discuss these ideas with your class.

3. What do all of the methods that allowed you to see a 3-D image have in common?

4. Reflect again on the earlier question regarding the reason for animals having two eyes—binocular vision. Do you agree with your original thinking? Why or why not? Use evidence you have gained from your activities to explain why animals have two eyes.

Extend

1. Have you ever used special glasses for viewing a 3-D movie? Share your experiences with the class.

2. One method for viewing a flat image in 3-D requires wearing glasses with one red and one blue lens. Your teacher will provide you with a picture and glasses. Look at the picture and describe what you see.

3. Closely observe the picture without the glasses. What do you notice?

4. Using colored pencils or markers, try to make your own picture that is 3-D when viewed with the glasses.

5. Explain how you think this 3-D system works. What do you see if you look at your drawing through the glasses with one eye closed? Then the other eye? How does each eye see a slightly different image?

Evaluate

Your teacher will show you a View-Master and a reel of pictures. Study them closely to determine how they work. Record in your journal how you see 3-D images with the View-Master.

PART 2

Engineering Materials

CHAPTER 5

PRODUCING PLASTIC ... FROM MILK?

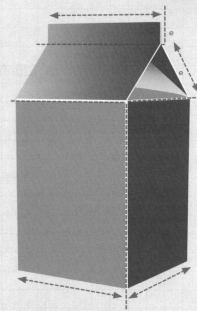

HOW MUCH PLASTIC have you handled today? Probably a great deal—perhaps a toothbrush, a shampoo bottle, a peanut butter jar, a plastic bag or two, the keys on your computer keyboard, and a pen or pencil. Just take a look at your car and see how much plastic it contains! In the 21st century, plastic is pervasive in our lives.

There are numerous types of plastics, most of which are derived from petroleum. Some are quite rigid and durable, including those used to make furniture, buttons, and many toys. Others, such as water bottles and food containers, are more pliable. Still other plastics can be drawn into fibers like nylon and rayon or into sheets to make wraps and bags. While there are many different types of plastics, they all have one thing in common—they are made of long chains of molecules called *polymers* that can be molded or shaped. In fact, the root word of plastic is from the Greek verb *plassein*, which means "to mold or shape" (Freinkel 2011). It is precisely for this reason that plastic is so common in our everyday lives—it can be made into a nearly unlimited number of shapes for a variety of uses.

The *Next Generation Science Standards (NGSS)* requires middle-level students to "[g]ather and make sense of information to describe that synthetic materials come from natural resources and impact society" (MS-PS1-3; NGSS Lead States 2013). The performance expectation is based on the disciplinary core idea of chemical reactions (PS1.B): "Substances react chemically in characteristic ways. In a chemical process, the atoms that make up the original substances are regrouped into different molecules, and these new substances have different properties from those of the reactants" (NGSS Lead States 2013). In this 5E learning-cycle lesson, students engage in the chemical-engineering challenge of determining the most economical way to increase the yield (the amount of plastic) of a chemical reaction. They will use milk and vinegar to produce casein, a type of plastic, and manipulate the amount of vinegar and the temperature of the milk.

This type of challenge is a major focus for many engineers who are responsible for manufacturing processes. The engineering practice of constructing explanations and designing solutions from the *NGSS* states that students "[u]ndertake a design project, engaging in the design cycle, to construct and/or implement a solution that meets specific design criteria and constraints" (MS-PS1-6; NGSS Lead States 2013). Note that while many students associate engineering with the design of objects such as cars, bridges, and computers (fewer realize that engineering is also related to the design of everyday objects such as pens and paper clips), it is less common for students to connect engineering with designing a process for making a product.

Historical Information

People have been using natural materials for millennia to make tools and other objects. Animal skins, elephant tusks, mud, clay, and many plant materials have been used. Patent books from the early 19th century, when people began to seek more durable materials, are "filled with inventions involving combinations of cork, sawdust, rubbers, and gums, even blood and milk protein, all designed to yield materials that had some of the qualities we now ascribe to plastic" (Freinkel 2011, p. 2). In 1867, the *New York Times* wrote of the possible extinction of elephants, since a million pounds of ivory were consumed each year, much of it to make billiard balls. A contest was held by a manufacturer of billiard balls to seek an alternative material, and in 1869, John Wesley Hyatt won the contest by discovering celluloid, which was made of cellulose from cotton and nitric acid. Celluloid billiard balls were never widely used because they, and the process to make them, were dangerous and flammable.

Casein, a plastic made from the proteins in milk, was used by ancient Egyptians in pigments for wall paintings. At the end of the 19th century, it was being produced as solid material. Many buttons and buckles were made from casein plastic during this time. Casein was also used to make knitting needles, fountain pens, combs, mirrors, and other items. Many may think of whiteboards as a relatively new invention, but at the end of the 19th century, when the use of paper by schoolchildren was considered too expensive, manufacturers were looking to replace slates with casein whiteboards (Plastics Historical Society 2011). As newer plastics were invented during and after World War II, the use of casein plastics diminished. Nonetheless, you can still purchase some casein products, such as buttons and faux-ivory piano keys.

Investigating Plastics (Teacher Background Information)

Materials

Students should wear indirectly vented chemical splash goggles, gloves, and thermal aprons during the labora-tory portions of this lesson. In the Explore phase, each group of four students will need four beakers, each with 150 ml of milk that has been heated to 49 °C. We recommend that you preheat the milk for the entire class. For the Extend portion, each pair of students will need an additional 150 ml of hot milk. Finally, for the Evaluate phase, each group of four students will require four more samples of 150 ml of milk at varying temperatures. Thus, for a class with six groups, you will need a total of 9 L of milk. You can expect to obtain about 25 samples of 150 ml from 1 gallon of milk. The fat content of the milk matters little in this investiga-tion, and the activity will also work with reconstituted powdered milk. To reduce the amount of milk used in the Explore and the Evaluate phases, each group could test only two samples and then combine class data.

For one group of students to complete all of the activities in each phase, 82 ml of plain white vinegar will be needed. Thus, for a class of six groups, you will need a little less than a half liter of vinegar (492 ml). The easiest way to filter the casein precipitate that is formed when vinegar is added to hot milk is to use a small square of cheesecloth (to fit over the beaker), which is available at grocery stores. Each group will also need a stirring rod and some means to measure the vinegar—we recommend metric measuring spoons. A slightly larger beaker is needed for each group to collect the decanted liquid. Newspaper or paper toweling is needed to absorb excess moisture from each sample. Cafeteria-type trays lined with wax paper can be used to dry samples overnight. Finally, each group of students will need a balance to find the mass of the dried samples of casein.

Small candy molds work well in the Extend phase. Molds with 10 cavities can be purchased at kitchen and craft stores for about a dollar; you may wish to cut them apart with scissors. If students want to poke holes in any of their plastic forms or to make beads, you will need to have some small wire or straightened paper clips available as well. Depending on what students decide to make, they may request additional supplies—such as safety pins to make pins, string to make bracelets, and food dye for coloring.

Engage

We begin this lesson by having students brainstorm plastic items they use on a daily basis. Since there are so many different types of plastic, it may be harder for them to develop a broad statement of what defines a plastic. In general, as already noted, plastics are materials that can be shaped or molded. Often, people think of plastics as flexible, though many are quite rigid. Another important property is that most plastics are chemically inert to many reactants. A major focus of this phase is for students to relate their prior knowledge to the chemical-engineering challenge of economically producing one type of plastic. Students will likely be surprised that plastic can be produced from the proteins in milk.

Explore

To begin the Explore phase, students need to observe the chemical reaction that precipitates the casein plastic from the hot milk. It is a striking reaction, and with 49°C milk and 15 ml of vinegar, the casein will develop (with gentle stirring) in 15 or 20 seconds (see Figure 5.1). Students may be able to pick up the white plastic mass and should note that the remaining liquid is no longer white like milk but a watery, grayish color, instead. This process is similar to the making of cottage cheese, the casein being the curds, and the watery liquid the whey. If students have difficulty separating the casein mass from the liquid, have them filter it using cheesecloth (see Figure 5.2). Instruct students to squeeze excess moisture out of the plastic and knead the plastic into a pancake shape (which will dry faster than a sphere) that they will put aside for massing another day when it has dried. Ask students why they think they should wait for the casein to dry before finding the mass (to only measure the mass of the casein and not the excess moisture). Students should have little difficulty observing the plasticlike characteristics of

FIGURE 5.1 15 ml of vinegar in 150 ml of hot (49°C) milk

FIGURE 5.2 Filtering casein plastic

FIGURE 5.3	Casein yields for 2, 5, and 10 ml of vinegar

TABLE 5.1	Data table for casein yield using 49°C milk

Amount of vinegar (ml)	Yield of dried casein (g)
2	1.4
3	2.1
5	5.9
15	8.3

the casein—it is a bit like putty, and as the moisture is removed, it can be readily molded into different shapes.

Students will now design a test to determine how to increase the yield of casein using the least amount of vinegar. It is recommended that students test three additional amounts of vinegar, with 15 ml being the largest (more vinegar will yield little additional casein). When approving student plans, make sure they keep the volume and temperature of the milk constant and change only the amount of vinegar. Again, students need to remove excess moisture and form the casein into pancake shapes that they set aside to dry (see Figure 5.3).

Explain

Students will find the mass of each casein sample the next day and construct a graph showing their results. Table 5.1 shows typical yields for 2, 3, 5, and 15 ml of vinegar. If all of the students did not use the same amounts of vinegar, consider combining all class data before constructing graphs. Students should discover that the yield of casein increases with increased vinegar up to a certain point where essentially all of the casein has been precipitated—you can see that the remaining liquid is no longer white at this point. It is the proteins in the milk that result in the white color; the remaining whey has little color. Thus, adding more vinegar does not result in any more casein.

Student graphs should show that the yield of casein increases to a point and then flattens to nearly horizontal, indicating no additional production with increased vinegar. Engineers use such a graph when they design the process for manufacturing casein so that excess vinegar is not used and essentially wasted. The graph shows a macroscopic pattern of how the yield of casein changes with varying amounts of vinegar. To address the crosscutting concept of patterns (NRC 2012), students need to understand what is happening at the microscopic level that corresponds to the graph—there is a finite number of protein molecules in the milk, and when they have been converted to the casein material, no more can be produced regardless of the amount of additional vinegar. Therefore, the graph flattens to become horizontal.

All plastics are composed of polymers, which are long chains of repeating units of molecules known as monomers. For an analogy to help understand polymers, consider the paper clips shown in Figures 5.4A and 5.4B. In Figure 5.4A, three paper clips have been chained together to form a model of a monomer. Note that in Figure 5.4B, three monomers are combined to represent the beginning of the formation of a polymer that in reality may contain thousands of monomer units.

FIGURE 5.4 Paper clip models

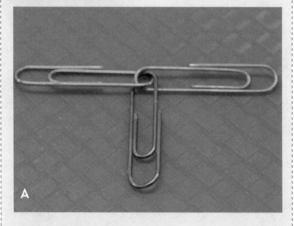

(A) Paper-clip monomer model
(B) Paper-clip polymer model

Extend

As noted earlier, one characteristic of plastic materials is their ability to be formed or molded into many different shapes. Students can form a variety of shapes using the casein plastic. Since the casein they have made previously has dried, they will need to make more for the Extend phase of the lesson. Students should again squeeze excess moisture out of the mass and then form it into a desired shape. Many may wish to make beads. After forming a bead shape, they can pierce it with a small wire or paper clip so the bead ultimately can be strung. This needs to be done while the casein is still moist. For more complex designs, candy molds work well for the small amount of casein students will use (see Figure 5.5). The casein can be colored by putting food dye into the milk before adding the vinegar. A messier alternative is to mix it into the casein while kneading it into a shape.

FIGURE 5.5 Molded items

One relatively simple plastic is polyethylene, which is often used to make plastic grocery bags. The monomer in this case is known as ethylene, a hydrocarbon (composed of the elements hydrogen and carbon) derived from petroleum with the chemical formula C_2H_4. The monomer unit can be represented like this:

$$-[H_2C-CH_2]-.$$

Polyethylene is formed when many thousands of these monomers combine much like the chain of paper clips. Casein plastic is a more complicated polymer that is formed by the reaction of milk proteins with an acid, in this case vinegar.

TABLE 5.2	Sample data for casein yield using 5 ml vinegar

Temperature of milk (°C)	Yield of casein (g)
36	0.96
49	5.90
54	8.60
64	8.10

Evaluate

To this point, students have investigated only one variable—the amount of vinegar. We purposely held the temperature constant at 49°C. The temperature of the milk is also an important factor in determining the yield of casein. Up to a point, the yield generally increases with temperature. As a performance assessment, students will plan and conduct an experiment to determine casein yield as a function of temperature (see sample data in Table 5.2). A critical issue is for students to recognize that the amount of vinegar must be kept constant. Depending on students' level of experience with experimental design, you may wish to have them identify dependent and independent variables and write a hypothesis.

Conclusion

Students may realize that, in actuality, the process of maximizing casein production is more complex than what we have addressed. Increasing the temperature also adds energy costs to the manufacturing process. A chemical engineer would have to weigh the cost of additional vinegar against the cost of the energy needed to increase the temperature of the milk during production. Many data points would have to be analyzed to determine the maximum yield at the lowest cost.

Like all nonnatural objects, the items we use every day have to be designed. The process for making them also has to be designed. The next time you look at a button, a plastic bead, or a toy, you may wonder about the engineering process that was used to create it.

References

Freinkel, S. 2011. A brief history of plastic's conquest of the world. *Scientific American*. May 29. *www.scientificamerican.com/article.cfm?id= a-brief-history-of-plastic-world-conquest*

National Research Council (NRC). 2012. *A framework for K–12 science education: Practices, crosscutting concepts, and core ideas.* Washington, DC: National Academies Press.

NGSS Lead States. 2013. *Next generation science standards.* Washington, DC: National Academies Press. *www. nextgenscience.org/next-generation-science-standards*

Plastics Historical Society. 2011. Casein. *http:// plastiquarian.com/?page_id=14228*

Resource

What is a polymer?—*www.pslc.ws/macrog/kidsmac/ basics.htm*

ACTIVITY WORKSHEET 5.1 — Investigating Plastics

Engage

1. Consider your activities of the past two days. Make a list of all the items you used that were made from plastic. Share your list with your classmates.

2. How do you know if an item is made out of plastic? What are some of the properties of plastic? From what materials do you think plastic is made?

3. In this activity, you will produce a plastic from ordinary cow's milk. Your engineering challenge will be to produce the most plastic using the fewest resources. How are the properties of milk and plastic different?

Explore

Safety note: Wear indirectly vented chemical-splash goggles, gloves, and a thermal apron while you conduct this part of the activity.

1. Stir 15 ml of vinegar into 150 ml of hot milk provided by your teacher, and record your observations of the changes that you see.

2. The solid matter in your mixture is a type of plastic called casein. Filter the mixture, if needed, to isolate the casein using the materials provided by your teacher.

3. Scrape all of the casein off the filter, roll the casein into a ball, flatten it into a pancake shape, and place it on a piece of newspaper or paper towel on a tray for it to dry overnight. When it is dry, you can find and record the mass of casein you have produced. Compare the properties of casein and plastic.

4. Now that you know how to make one type of plastic, imagine that you are a chemical engineer whose job it is to devise a method to produce as much casein plastic as possible using the least amount of vinegar. In other words, does the amount of vinegar affect the amount of casein produced? In planning your procedure, consider what will happen if you vary the amount of vinegar. What factors need to be kept constant in your procedure? Use an "if ... then" statement to make a prediction of what you think will happen.

5. After you have written down your procedure, get your teacher's approval before proceeding.

6. Conduct your experiment and record your data in a table. Be sure to label and set aside each casein sample to dry before you measure its mass.

Explain

1. Complete your data table by massing each sample after it has dried.

2. Construct a graph showing the amount of vinegar used and the mass of casein produced.

3. What happens to the yield of casein as you increase the amount of vinegar? Does it support the prediction you made? Use data to support your answer.

4. How would your graph help a chemical engineer determine how to produce casein plastic in the most economical way?

Extend

Safety note: Wear indirectly vented chemical-splash goggles, gloves, and a thermal apron while you conduct this part of the activity.

One characteristic of plastic is that it is easily formed or molded into different shapes. At one time, many buttons, buckles, and small jewelry items were made from casein plastic. Using the materials your teacher provides, design a product out of casein. You might want to make beads for a bracelet or use a candy mold to make a pendant. Be a creative engineer and make something you will enjoy!

Evaluate

You have tested one variable, the amount of vinegar, related to the process of making casein plastic from milk. Record other possible variables. Consider the temperature of the milk. Design a test to see if the temperature of the milk affects the yield of casein produced. Record and carry out your plan after your teacher has approved it. Compare the casein yield if you vary the temperature versus if you vary the amount of vinegar.

CHAPTER 6

IF IT'S ENGINEERED, IS IT WOOD?

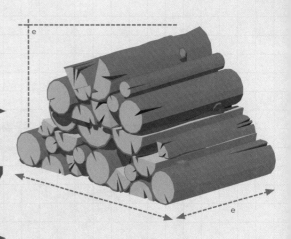

CONSIDER ALL THE "wooden" items around you. Which ones are made from natural wooden planks, and which are made from manufactured wood products, such as sawdust, chips, or laminates? Take a closer look. For example, by looking at a piece of wooden furniture, can you determine if it is made from wood products or timber sawed into planks or boards? What other wooden items do you see—doors, bookshelves, coatracks, pencils, and assorted knickknacks? Are any made of simply sawed timber or what we might call *real wood*? How can you identify manufactured or engineered wood?

Manufactured or engineered wood products are made "by bonding together wood strands, veneers, lumber or other forms of wood fiber" (APA—The Engineered Wood Association 2014), which results in a product with desirable characteristics (and often efficient use of natural resources). You are likely familiar with many types of engineered wood products, such as plywood, laminate flooring, and particleboard.

In this 5E learning-cycle lesson, students will explore one aspect of manufactured wood, namely, strength. Students will compare the deflection of a beam made of layers of balsa wood that have been glued together to an equal number of layers that have not been glued. Students will also analyze a number of common, engineered wood products to determine how they are constructed and used. The *Next Generation Science Standards* include the following for the middle-level student: "Gather and make sense of information to describe that synthetic materials come from natural resources and impact society" (MS-PS1-3; NGSS Lead States 2013). In this lesson, students will test how strength can be increased with engineered wood and also learn that engineered woods have evolved from simple plywood and particleboards to a variety of more complex veneers and laminates. One disciplinary core idea for engineering design (MS-ETS1) at middle level suggests that students understand "[t]he iterative process of testing the most promising solutions and modifying what is proposed on the basis of the test results leads to greater refinement and ultimately to an optimal solution" (MS-ETS1.C; NGSS Lead States 2013). Over the course of the past century or so, starting with plywood, this is exactly what has occurred, as more and more engineered wood products have been developed to meet the requirements of varying consumer needs.

Historical Information

While most of the history of engineered wood products is relatively recent, evidence of laminated wood has been found in Egyptian tombs, and the Chinese glued

FIGURE 6.1 Plastic skateboard

FIGURE 6.2 Wooden skateboard

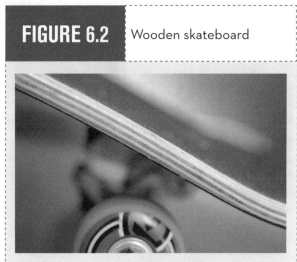

shaved wood pieces together to make furniture about 1,000 years ago. The first patent for plywood was issued in 1865 to John K. Mayo, although little was done with the idea at that time. During the 1905 World's Fair in Portland, Oregon, Gustav Carlson displayed wood made from three laminated sheets. These panels caught the interest of door makers. Plywood became an extremely popular material for the construction of furniture, running boards for automobiles, and many other applications (APA—The Engineered Wood Association 2014).

In the 1930s, waterproof adhesives were developed, allowing outdoor construction uses. Plywood could now be used for floors, roofs, and sheathing for walls. During World War II, the navy made substantial use of plywood boats, including D-Day landing craft, as well as patrol boats in the Pacific.

Oriented strand board (OSB), which is made of glued wood strands or fibers, was introduced in the late 1970s. It has similar characteristics to plywood but is less expensive to produce. Other building materials have been developed since the introduction of OSB, including OSB lumber and joists (APA—The Engineered Wood Association 2014).

Investigating Engineered Wood (Teacher Background Information)

Materials

For the Engage phase of this lesson, try to procure some skateboards for students to examine. Inexpensive skateboards tend to be made of either wood or plastic, but higher-quality skateboards are usually made of laminated wood, typically maple (see Figures 6.1 and 6.2). Have at least one of each. (You can also download skateboard photos at *www.nsta.org/more-engineering*.) (*Note:* If students bring skateboards to school for use in this part of the activity, they must obey all school rules while the boards are in the building. Some schools might even require parental permission for the student to bring in the skateboard. Students should not play with the boards on school property.)

In the Explore phase, students need to wear indirectly vented chemical-splash goggles and cover their tables with newspaper. Each group of four students will need eight pieces of $1.5 \times 76.2 \times 457.0$ mm ($^1/_{16} \times 3 \times 18$ in.) balsa wood, which should be prepared prior to class: three individual pieces, one board where two pieces have been glued together with white glue, and one where three pieces have been glued.

You will need to either clamp or apply a weight to the glued boards and let them dry overnight. Mark a line 10 cm from one end of each board. The balsa boards are sold in pieces twice this length for less than $2 and can easily be cut with a utility knife. The balsa wood can be reused for each class and saved for subsequent lessons. In addition, each group will need a 473 ml (16 oz.) foam cup, a 250 ml graduated cylinder, a container of water about 2 L in size, and a textbook that is about 3.5 to 4.0 cm (1.4 to 1.6 in.) thick.

Because of their combined strength, you will need to devise an alternative method of testing the three plies glued together. We attached a large paper clip through the end of the three-ply board and hooked on a plastic bucket. Because the end of the balsa wood now extends beyond the table, we used a ruler taped to the table to see when the board deflected to the plane of the table. Thus, you will need one bucketlike container, a ruler, and a piece of duct tape for each group.

For the Extend phase, gather some samples of engineered wood products for students to use: plywood, fiberboard, particleboard furniture, OSB, laminated flooring, and so forth. Ideally, each group should have three different samples. You may find these as scraps in local lumber outlets, in school wood shops, or from friends and neighbors. Each student should have a hand lens of at least 3× magnification. If you are unable to gather these samples, you can find close-up photos of engineered wood products on the internet.

Engage

Many of the best skateboards are constructed from about seven layers of laminated wood, usually sugar maple. Others are made from a variety of synthetic materials. However, many accomplished riders prefer the combination of characteristics found in wooden boards—strength, flexibility, feel, and response (Wanner 2014). It is also easier to form wood laminates into curved shapes. Students are better able to see the laminates on wooden skateboards that have not been painted. Students will notice that in addition to the

material being different, the shapes, surface, wheels, and amount of curvature may all vary to some degree.

Focus students' attention on the edges of the laminated wood so they notice the layers. To determine students' prior knowledge of engineered wood products, you can ask how they think the skateboard was made. Some possible answers might include that it was made from a single block of wood or, if students notice the lines on the edge of the skateboard, from thin layers of wood. Other students may incorrectly interpret the lines as tree rings, somehow related to the age of the tree. While it is unlikely that students will realize any of the above considerations, they may suggest that the laminates are somehow stronger.

Explore

Safety note: In case of the unlikely event that the balsa wood splinters, students must wear indirectly vented chemical splash goggles and cover tables with newspaper throughout this exploration.

Depending on the experience of your students, you may wish to have them develop their own procedures for testing the questions from the Engage phase: Does the strength of a board change as more layers of wood are added? Will gluing the layers together affect the overall strength of the board? We have provided one method in the student Activity Worksheet (see Figure 6.3, p. 42). To measure the strength of the three plies glued together, a much larger container than the 16 oz. foam cup noted in the "Materials" section is needed. It is likely that this board will support somewhere between 1 and 2 L of water. Thus, some sort of strong, bucketlike container is required. To reduce the possibility of spillage, we suspended a bucket with a paper clip (see Figure 6.4, p. 42).

Students will collect their data in milliliters of water but will need to convert these data to mass units. Because 1 ml of water has a mass of 1 g, the volumes recorded, plus the mass of the container, can be used as a measure of the strength of the various balsa boards (in grams). The mass of the container must

FIGURE 6.3 | Setup and results of cantilevered balsa boards

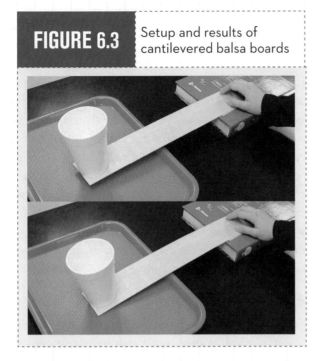

FIGURE 6.4 | Using large container to test three glued boards

be added to the mass of the water. See our sample in Table 6.1.

Explain

As students analyze their data, they should have no difficulty inferring that as the number of plies is increased, both the volume and the mass of the water that can be supported will also increase. This indicates that the strength of the balsa wood boards increases as the number of plies increases. They should note that gluing the plies together results in a dramatic increase in the strength of the boards. While you can expect variation among the data collected by individual groups, the groups should arrive at similar conclusions.

Gluing the plies together prevents them from slipping over one another, resulting in a more rigid structure that can resist deflection when weighted. Like the balsa wood boards, plywood is also made up of a number of plies—usually ranging from five to nine layers. The grain of each ply runs in different directions, which also increases the strength. Finally, the adhesive used to manufacture plywood is generally stronger than the fibers of the wood it is gluing together. For any given mass, plywood is stronger than steel in bending strength (HPVA 2014).

Extend

Help students appreciate that the samples of glued balsa wood boards tested were examples of engineered wood products. The gluing of the boards results in a product that is more resistant to bending and far more rigid than the natural, unglued boards.

Students should note that the engineered wood products they examined are made up of either layers or small pieces of wood that are glued together. They differ in the size of the pieces that are glued together, ranging from sheets to fibers to sawdust. For example, plywood (see Figure 6.5) is made up of individual, thin sheets of wood that are glued together. Figure 6.6 shows OSB, which is made of small fibers or flakes of wood that are glued together. Both plywood and OSB are often used in building construction for roofs, walls, and flooring.

TABLE 6.1 Sample data

Layers	Unglued (ml)				Mass (g)		Glued (ml)				Mass (g)	
	Trial 1	Trial 2	Trial 3	Average	Mass of container	Mass of water and container	Trial 1	Trial 2	Trial 3	Average	Mass of container	Mass of water and container
1 ply	40	44	44	42.6	4	46.6	—	—	—	—	—	—
2 plies	74	73.5	71	72.8	4	76.8	226	238	231	232	4	236
3 plies	102	97.5	98	99.2	4	103.2	1,300	1,282	1,326	1,303	104	1,407

FIGURE 6.5 Engineered wood—plywood

FIGURE 6.6 Engineered wood—OSB

FIGURE 6.7 Engineered wood—laminate flooring

One common use of engineered wood is for flooring. Figure 6.7 shows a typical example of a laminated wood flooring product. The bottom layer, or *underlayment*, is for sound absorption and stability. The core layer is usually made of high-density fiberboard (HDF), which is made of very small wood fibers glued together. The top is a thin sheet of paper printed with either wood grain or some other pattern and then covered with a clear resin designed for durability. Thus, the only wood in laminate wood floors is the core HDF.

HDF is also used for furniture construction and other purposes. Particleboard is similar in construction to HDF, except it is made of waste-wood chips and saw-dust that are glued together. It is therefore less expensive, but also weaker, than HDF. It is used to make inexpensive furniture—often of the variety that is purchased in a box and put together by the end consumer.

Evaluate

Students will use their data to construct a graph from which they can compare the strength of the glued and unglued balsa wood boards (see Table 6.1, p. 43). Students should conclude that two glued boards are significantly stronger than both two and three unglued boards. Therefore, gluing increases strength more than adding additional layers. Thus, by gluing, one can construct a stronger board using less material.

Conclusion

This simple exploration includes science, technology, engineering, and math disciplines in one learning-cycle lesson. The science content relates to recognizing that synthetic materials can be derived from natural resources. The resulting technology is the glued boards that exhibit excellent strength characteristics. Indeed, there exists a plethora of engineered wood products with desirable characteristics that can be manufactured economically from natural wood or, in many cases, wood-waste products. The engineering consists of students testing one possible solution to increasing the strength of a board—gluing a number of plies together. This refinement leads to an optimal solution for increasing the strength of a board but keeping the overall amount of wood needed to a minimum. Finally, students need to apply mathematics to determine which added more strength, adding additional plies or gluing them together.

In this lesson, students learned that engineered wood is made up of glues or resins that hold together natural wood sheets, fibers, dust, or particles. To answer the question posed in the title of this article, engineered wood products are indeed made up of, among other materials, at least some wood.

References

APA—The Engineered Wood Association. 2014. Wood University glossary. *www.wooduniversity.org/glossary*

Hardwood Plywood & Veneer Association (HPVA). 2014. Advantages. *www.hpva.org/hardwood-plywood/advantages*

NGSS Lead States. 2013. *Next Generation Science Standards: For states, by states.* Washington, DC: National Academies Press. *www.nextgenscience.org/next-generation-science-standards*

Wanner, N. 2014. Skateboard science: The science and art of skateboard design. *Exploratorium. www.exploratorium.edu/skateboarding/skatedesign.html*

Resources

Moyer, R. H., and J. E. Bishop. 1986. *General science.* Columbus, OH: Charles E. Merrill. *Note:* The balsa wood activity was first published on page 175 in this book.

Vassiliev, T., P. Bernhardt, and D. Neivand. 2013. Innovative composite research modeled in the middle school classroom. *Science Scope* 37 (1): 42–52.

World Floor Covering Association. How laminate flooring is made. *www.wfca.org/Pages/How-Laminate-Flooring-Is-Made.aspx*

ACTIVITY WORKSHEET 6.1 — Investigating Engineered Wood Products

Engage

1. Think about a skateboard. What material do you think is used to make the deck?

2. Observe the illustration of the two skateboards your teacher has provided. What materials do you think were used to make the deck of each skateboard? On what observations do you base your inferences?

3. How are the two decks similar? Different?

4. The wooden deck of one skateboard is made of layers of wood that have been glued and then pressed together. In this exploration, you will investigate if the strength of a board changes as more layers of wood are added. Also, does gluing the layers together affect the overall strength of the board?

Explore

Safety note: Wear indirectly vented chemical splash goggles and cover tables and the floor with newspaper (in case of spills). Keep hands and feet away from the bucket during testing.

1. In this exploration, you will build a model of a skateboard deck out of balsa wood and test the strength of your model when it is made of one, two, and three layers of balsa wood. You will also test its strength when the layers of balsa wood are glued together.

2. To test the strength, use a setup that looks something like a diving board. Overlap 10 cm of one end of the board on a textbook and then place a foam cup on the other end of the board.

3. While a classmate holds the end of the board overlapping the book, slowly add water to the cup until the balsa wood board bends sufficiently to just touch the table. Pour the water from the cup into a graduated cylinder and record the amount in a table like the one shown below. Repeat for two additional trials.

4. Repeat the above procedure with two layers of balsa wood (unglued) and then three (unglued) layers.

5. Repeat the procedure again for the two layers of balsa wood that are glued together.

6. To test the three layers glued together, you may need to alter your procedure. Discuss your plan with your teacher before proceeding.

7. Find the averages for each group of three trials.

8. Find the mass of your containers and record these in the table. Recall that the mass of 1 ml of water is 1 g. Determine the total mass that caused the balsa wood boards to bend by adding the mass

Layers	Unglued (ml)				Mass (g)		Glued (mL)				Mass (g)	
	Trial 1	Trial 2	Trial 3	Average	Mass of container	Mass of water and container	Trial 1	Trial 2	Trial 3	Average	Mass of container	Mass of water and container
1 ply							—	—	—	—	—	—
2 plies												
3 plies												

of the container to the average mass of the water. Record this information in the table.

Explain

1. Look at the data in your table and determine how the number of layers affects the rigidity of the balsa wood board. What is the relationship between the mass of water supported and the number of layers of balsa wood? What is the relationship between the number of plies of balsa wood and the strength of the balsa wood boards?

2. What affect does gluing the layers together have on the strength?

3. Share your results with the other groups in your class to see how the data compare. How do each group's conclusions compare?

Extend

1. In the Explore stage, you tested samples of engineered wood products that were made from glued layers of balsa wood. How does gluing the layers together improve the usefulness of the board?

2. Look at the samples of other engineered wood products that your teacher has provided. Use a hand lens to observe the characteristics of each item. What kinds of materials do you think were used to produce them? Make a sketch of each product and label it to show the different parts.

3. How are the various engineered wood products similar? Different?

4. Find additional examples of engineered wood products in your school or at home. Make sketches of each one showing the different parts.

Evaluate

There were two independent variables in this exploration: the number of plies of balsa wood and whether the plies were glued together. You have collected data on how each of these variables relates to the strength of the resulting balsa wood boards. Which increases the strength more, adding additional plies or gluing the plies together? Construct a graph using the averages from your exploration data to confirm your answer.

CHAPTER 7

UV OR NOT UV?
THAT IS A QUESTION
FOR YOUR SUNGLASSES

HOW MANY PAIRS of sunglasses do you own? How do you decide which type to purchase? Choices range from dollar-store basics to high-end designer shades costing hundreds of dollars. Some people buy sunglasses for special purposes—for example, skiing, boating, or other outdoor recreational activities. Others own many pairs of inexpensive sunglasses to have some on hand at all times. You might also purchase prescription sunglasses. In addition, more and more people consider the health benefits of sunglasses, namely, how well they filter damaging ultraviolet (UV) light. According to the U.S. Environmental Protection Agency (2010, p. 1), "Long-term exposure to UV radiation can lead to cataracts, skin cancer around the eyelids, and other eye disorders."

In this 5E learning-cycle lesson, students use UV-sensitive beads to test different sunglasses' lenses to determine their ability to filter UV light. UV-sensitive beads are white but change color when exposed to UV light and return to white when the source of UV light is removed. In the *Next Generation Science Standards*, students should recognize that "[w]hen light shines on an object, it is reflected, absorbed, or transmitted through the object, depending on the object's material and the frequency (color) of the light" (MS-PS4-2; NGSS Lead States 2013). This is essentially what

sunglasses, like virtually all filters, do. Some of the light (including the UV portion) that strikes the lenses of sunglasses is reflected, some is absorbed (most of the UV portion), and the rest is transmitted through to the wearer's eyes. Sunglasses that are effective in protecting the eyes from UV light do so mostly by absorbing these harmful rays. UV radiation is that portion of the spectrum with wavelengths shorter than visible light. The longest wavelength of visible light (red) is 700 nanometers (nm), and the shortest is violet at 400 nm. Most of the UV radiation that reaches the Earth has wavelengths between 290 and 400 nm. This range includes what is referred to as UVA (320–400 nm) and UVB (290–320 nm) radiation. Most of the shorter wavelengths are filtered by the atmosphere and do not reach the Earth's surface (Skin Cancer Foundation 2016).

As students test different lenses, they will "[e]valuate competing design solutions using a systematic process to determine how well they meet the criteria and constraints of the problem" (MS-ETS1-2; NGSS Lead States 2013). In the United States, there are voluntary standards for sunglasses administrated by the American National Standards Institute (ANSI; ANSI 2015). The criteria in the ANSI standards require that sunglasses filter essentially 99% of the UV light below

48

wavelengths of 400 nm, allow colors to be perceived accurately, and be impact resistant.

Historical Information

As early as the 1200s, the Inuit of North America cut thin slits into walrus ivory to make a type of snow goggle to reduce the Sun's glare off of the snow and ice (Canadian Museum of History, n.d.). One of the earliest references to sunglasses appears in 1750, when Englishman James Ayscough noted that when glasses were "a little ting'd with Blue, it takes off the glaring Light from the Paper, and renders every Object so easy and pleasant" (Ayscough 1750, p. 13). Not until the 1920s, however, did sunglasses become popular with the general public. In 1929, Sam Foster sold inexpensive sunglasses on the boardwalk in Atlantic City, a popular tourist destination (Foster Grant 2013), and they quickly became a national fad (Life 1938). In the 1930s, Edwin Land (the same person who developed instant film cameras) invented sheets for polarizing light, an invention that was then used in sunglasses to reduce reflective glare (Rowland Institute at Harvard, n.d.). UV-protective sunglasses are a spin-off of NASA research at the Jet Propulsion Laboratory from the 1980s, where scientists studying birds of prey discovered that the birds produced an oil that protected their eyes from UV light. This research was first applied to the design of equipment for shielding welders' eyes and, ultimately, sunglasses (NASA 2011).

Investigating Sunglasses (Teacher Background Information)

Materials

Throughout this lesson, you will use UV beads—small plastic beads that have been treated so that they change color when exposed to UV light. In the absence of UV light, they are white in color but are available in numerous colors in the presence of UV light. We have used beads that change from white to purple. Similar beads are available at science supply companies for about $9 for 1,400 beads. After numerous cycles, they no longer completely return to white but remain milky gray. Be aware that there are also beads that glow different colors when exposed to UV light. For this activity, you want to be sure to obtain UV beads, not UV glow beads. Students will need to keep the beads away from UV light until they are ready for testing.

In the Engage phase, each student will need one bead, which will be sufficient for all of the other parts of the lesson. You may wish to start each class with fresh beads. In the Explore phase, each group of four students will need three different sunglass lenses. These can be obtained by taking apart old sunglasses or by purchasing them at a dollar store. One lens must not meet the ANSI standards noted earlier. You should be able to find these at dollar or party stores labeled as "fashion" or "novelty" glasses. The lenses can be reused from one class to the next. For students' lens tests, you will need to cut apart egg cartons. Each student group will need four individual egg holders; therefore, you will need one egg carton for every three groups of students—if you have five classes with six groups in each, you would need a total of 10 egg cartons. To tape the lenses to the egg holder, each group will need approximately 50 cm of duct tape. In addition to at least one pair of sunglasses, the same materials will be reused in the Extend phase of the lesson. Several small plastic bottles are needed for the Evaluate phase of the lesson. These can include prescription bottles, over-the-counter medication bottles, and small plastic vials.

Ideally, you need a bright window to test the various lenses. Most windows do not filter all of the Sun's UV light. Alternatively, you can take students outside. The activity will work on a cloudy day, but more time will be required for the beads to change color.

Engage

Begin this lesson by initiating a discussion of different ways people protect themselves from the Sun. We can classify damaging aspects of the Sun's radiation into three different categories: (1) People must protect themselves from the warmth of the infrared portion of the Sun's spectrum so as to not become dehydrated or suffer heat stroke. (2) The visible part of sunlight is very bright and can damage the eyes if looked at directly. (3) Perhaps the most widespread concern for people is the UV radiation found in sunlight, which can lead to serious skin and eye damage.

Students will likely suggest wearing hats, sunscreen, and sunglasses, and, perhaps, limiting the amount of time spent in direct sunlight or keeping skin covered with long sleeves and long pants. This discussion may also reveal some common misconceptions students have, such as that getting a good tan can protect you from the Sun or that people of color cannot get skin cancer (Skin Cancer Foundation 2013).

To understand that the UV bead is an indicator of UV radiation, students should begin by placing their bead in a windowsill (see Figure 7.1), which emphasizes that the bead will cycle from white to colored and back to white in the presence or absence of UV light.

At this point in the lesson, students should be aware that sunlight contains UV radiation that is potentially harmful to humans. This lesson focuses specifically on learning how to test sunglasses to see if they are able to protect the eyes from this type of radiation.

| FIGURE 7.1 | UV beads change color on a sunny windowsill |

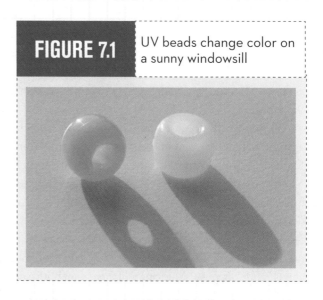

Explore

Working in groups of four, students will test several different sunglass lenses. You may wish to have them design their own testing procedure or use the one provided in the student Activity Worksheet. While students are preparing their egg-carton testing apparatus (see Figures 7.2 and 7.3), they should keep the beads away from UV light, including classroom fluorescent

| FIGURE 7.2 | Bead taped to inside of egg holder |

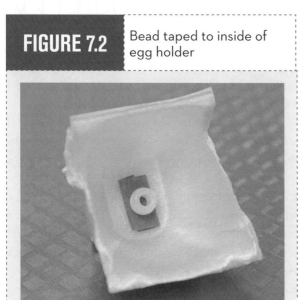

| FIGURE 7.3 | Bead in egg holder with sunglass lens |

lighting, which may also cause a nominal tinting of the beads. The easiest way to do this may be to simply cover the beads with a sheet of paper or, once the bead is attached to the egg holder, turn it upside down on the table.

With a piece of folded tape, students will attach a bead to the bottom of one egg holder (as shown in Figure 7.2, p. 49) to prevent it from rolling into a corner where it may be in a shadow. A second bead is secured in the same manner in another egg holder and then the opening is covered with a lens from a pair of sunglasses. Students should seal around the lens with tape so that no light enters the second egg holder except through the lens (Figure 7.3, p. 49). Students will then point both egg holders directly toward the sunlight through a classroom window. On a bright, sunny day, students should see the uncovered bead change color in less than a minute. They should also deduce from this that the other bead has had enough time to change. If it has not, the lens must be filtering most of the UV light. Students will repeat this process for each lens. You should be able to conduct this testing on a cloudy day, although it will take longer for the beads to change color. Again, students can use the uncovered bead as a testing time indicator. *Safety note:* Warn students never to look directly at the Sun.

Explain

Students will find that most recently manufactured sunglasses work equally well at filtering UV light, as required by ANSI standards. Any glasses that do not meet these standards will allow some UV light to pass through the lens, resulting in the bead's color change. For a given amount of time, bead color variation to a darker shade indicates that more UV light has passed through the lens. Thus, the best lenses will allow virtually no change in the color of the bead and therefore provide the most eye protection.

Students may ask if polarized sunglasses protect the eyes from UV light. Polarized lenses are designed to cut glare that reflects off of horizontal surfaces rather than to filter UV light. However, many polarized sunglasses

FIGURE 7.4 Child wearing wraparound sunglasses

are also treated for the filtration of UV rays (EyeCare America 2007).

Extend

Discuss with students how sunlight might enter their eyes even if they are wearing sunglasses, because, unlike in their testing, sunglasses are not taped to their faces. Students can demonstrate this design flaw by using a bead, egg holder, and lens that has not been sealed with tape. Students will observe that this bead changes color, indicating the presence of UV light.

The engineering challenge is to design a pair of sunglasses that will address these weaknesses and thus decrease the amount of UV light that enters the wearer's eyes (see Figures 7.4 and 7.5). All engineering designs have criteria to be addressed and constraints to be mitigated, which might lead to multiple solutions to a design challenge. For example, the child in Figure 7.4 is wearing a pair of wraparound sunglasses that block light from entering her eyes from the side. However, this design significantly limits her peripheral vision. Note how this constraint has been addressed by the

FIGURE 7.5 Assorted wraparound sunglasses

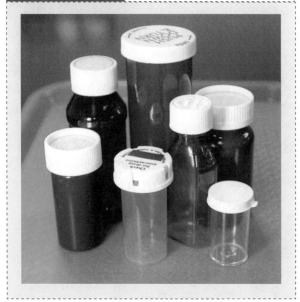

FIGURE 7.6 Small plastic bottles

design that includes a side lens in the adult glasses pictured in Figure 7.5.

Have students make sketches of their ideas, present their designs to the class, and explain how their designs provide additional protection.

Evaluate

As a performance assessment, students plan and conduct a test to determine if plastic medicine bottles filter UV light. If you gather used bottles (which come in a variety of colors), they should be thoroughly washed. You may also purchase this type of bottle from nearly any pharmacy. Students should now be able to design and conduct a similar test to the one previously described for sunglass lenses. Nearly all prescription bottles will filter UV light, so include additional small plastic vials as well (see Figure 7.6).

Conclusion

While there is much pressure for middle-level students to focus on fashion and design, ideally this engineering lesson will open their eyes to the importance of

additional criteria when choosing sunglasses, including the filtration of UV light.

References

American National Standards Institute (ANSI). 2015. Ophthalmics—Nonprescription sunglass and fashion eyewear requirements (ANSI Z80.3-2015). *webstore. ansi.org/RecordDetail.aspx?sku=ANSI+Z80.3-2015*

Ayscough, J. 1750. *A short account of the nature and use of spectacles. In which is recommended, a kind of glass for spectacles, preferable to any hitherto made use of for that purpose. http://quod.lib.umich.edu/e/ecco/0 04798249.0001.000?rgn=main;view=fulltext*

Canadian Museum of History. n. d. Snow goggles. *http:// collections.civilisations.ca/public/pages/cmccpublic/ alt-emupublic/Display.php?irn=855927*

EyeCare America. 2007. Sunglasses. *www.aao.org/ eyecare/tmp/sunglasses.cfm*

Foster Grant. 2013. About us: Origins of sunglasses. *www.fostergrant.co.uk/about.*

Life. 1938. Dark glasses are new fad for wear on city streets. *Life,* May 30. *http://books.google.com/books?id=3EoEAAAAMBAJ&lpg=PP1&pg=PA31#v=onepage&q&f=true*

National Aeronautics and Space Administration (NASA). 2011. Ultraviolet-blocking lenses protect, enhance vision. NASA Spinoff. *http://spinoff.nasa.gov/Spinoff2010/hm_3.html*

NGSS Lead States. 2013. *Next Generation Science Standards: For states, by states.* Washington, DC: National Academies Press. *www.nextgenscience.org/next-generation-science-standards*

Rowland Institute at Harvard. n.d. Edwin H. Land. *www2.rowland.harvard.edu/book/edwin-h-land*

Skin Cancer Foundation. 2013. The Skin Cancer Foundation busts myths surrounding Vitamin D and Sun exposure. *www.skincancer.org/media-and-press/press-release-2013/vitamin-d*

Skin Cancer Foundation. 2016. UVA & UVB. *www.skincancer.org/prevention/uva-and-uvb#*

U.S. Environmental Protection Agency. 2010. Prevent eye damage: Protect yourself from UV radiation. *www2.epa.gov/sites/production/files/documents/eyedamage.pdf*

Resource

Solar radiation and human health: Too much Sun is dangerous. WHO Information Fact Sheets. *www.who.int/uv/resources/fact/en/fs227toomuchsun.pdf*

ACTIVITY WORKSHEET 7.1 | Investigating Sunglasses

Engage

1. What are some different ways that people protect themselves from the Sun? Share these ideas with your classmates.

2. Your teacher will provide you with a white plastic bead that is sensitive to certain types of light. Place your bead in a sunny window for about a minute, less if you notice a change in the bead's color. Record your observations.

3. Remove the bead from the window and hold it in your hand so that it is no longer exposed to any light for about two minutes. What happens to the bead after it has been removed from the sunlight near the window? Record your observations. These beads are sensitive to the ultraviolet (UV) light that is a part of sunlight. UV light, however, is invisible to your eyes. Therefore, the change in color of the beads indicates the presence of UV light.

4. UV light is harmful to your eyes. This is one of the reasons why people wear sunglasses. Are all sunglasses equally effective in protecting your eyes from this type of light? In this exploration, you will use the UV beads to test the lenses from different types of sunglasses.

Explore

1. You need to construct a testing apparatus so that the only light that reaches the UV bead must pass through the lens. You also need to make sure that the beads do not get exposed to UV light until you are ready to conduct a test. One method is to place a UV bead in one of the single egg holders from an egg carton. Fold over a piece of tape to secure the bead to the center of the egg holder. Cover the opening with the lens from one of the sunglasses you wish to test and then use duct tape to seal around the lens.

2. Tape a second UV bead into an egg holder without a lens attached and then hold each in a sunny window with both pointing directly at the Sun for one to two minutes. Do not look at the Sun directly while doing this part of the activity.

3. Return to your work area away from the bright window and quickly unseal the egg holder and compare the color of the two beads. Record your observations in a table like the one shown below and repeat for the other lenses. Be sure to start each test with a fresh white UV bead.

UV Bead Color Observations

Lens	Egg holder with no lens	Egg holder with lens
Lens # 1		
Lens # 2		
Lens # 3		

Explain

1. Which lens resulted in the darkest color change of the bead? The least?

2. Share your results with your classmates.

3. What does the amount of color change mean? Would a darker or lighter bead indicate a better lens to protect your eyes from UV light? Explain.

4. How do the lenses compare? Explain.

Extend

1. You know that UV light is harmful to your eyes and that some sunglasses are able to filter most

of the UV light that passes through them. How might UV light enter your eyes even if you are wearing sunglasses? Record your ideas.

2. Test a lens in the same way as above but without sealing it to the egg holder with tape. You will need to hold the lens in place while pointing it at the Sun. After one or two minutes, return to your work area and observe the UV bead and record your observations.

3. Compare these results with the testing done in the Explore stage, when you sealed the lens to the egg holder. Which test better represents how people actually wear sunglasses?

4. Your engineering challenge is to design a pair of sunglasses that will decrease the amount of UV light entering the wearer's eyes.

5. Examine the glasses your teacher provided to determine their design weaknesses that allow UV light to enter the wearer's eyes. Brainstorm ways the sunglasses could be modified to address these design weaknesses. Make a sketch of your improved sunglasses, labeling your changes. Explain how your design better protects the eyes from UV light. Present your design to your classmates.

Evaluate

There are many other things besides your eyes that can be damaged by exposure to UV light. One example is medication that is often sold in small plastic bottles. Design a procedure to test if different types of small bottles filter UV light. After your teacher has approved your plan, carry out your testing. Record your results and explain your findings.

CHAPTER 8

WHY THE STATUE OF LIBERTY IS GREEN
Coatings, Corrosion, and Patina

DO YOU HAVE a jar of pennies somewhere? Maybe a change purse that is a bit too full? Or do you leave pennies in the little bowl on the counter at the store? While in one sense pennies are all alike, you may notice differences among them, especially in their color. Some are new and shiny, some are dull, some are brown, and a few show a little bit of green. Have you ever wondered why these variations exist?

You may have noticed other copper items that exhibit a green *patina*, or film: roofs and gutters, pipes, and even the Statue of Liberty. Originally, the copper outer surface of the Statue of Liberty looked more like a new penny, but over time it corroded to the green color we see today. Some people wonder if the statue is actually painted green. In 1906, Congress proposed painting the statue, but with copper-colored paint, to restore its original look (CDA 2014). In 1986, prior to the 100-year celebration of the statue's arrival in New York as a gift from France, it was completely cleaned and restored to its brighter copper color. In a few years, however, it once again took on its familiar green patina. (See the "Resources" section at the end of this chapter for videos about the Statue of Liberty's history and 1986 restoration.)

In this 5E learning-cycle lesson, students test different types of coatings on pennies to observe how the coatings affect the amount of corrosion produced when the penny is placed in a moist environment and a moist, acidic environment. This activity addresses performance expectation MSPS1-2 of the *Next Generation Science Standards* (*NGSS*) MS-PS1: Matter and Its Interactions, which asks students to "[a]nalyze and interpret data on the properties of substances before and after the substances interact to determine if a chemical reaction has occurred" (NGSS Lead States 2013). In this lesson, students note the degree of patina formed on each penny as an indicator of the extent of the chemical reaction. One *NGSS* engineering expectation for this level requires students to "[e]valuate competing design solutions using a systematic process to determine how well they meet the criteria and constraints of the problem" (performance expectation MS-ETS1-2 from standard MS-ETS1: Engineering Design; NGSS Lead States 2013). In this activity, the design solutions are the various coatings—such as oil, paint, and polish—being tested to determine the amount of protection each provides.

CHAPTER 8

Historical Information

A major type of coating is, of course, paint. Paints date to the very beginning of recorded history. The first paints, made from natural pigments such as animal fats and eggs, were developed by early humans 25,000 to 30,000 years ago and used in cave drawings found in Europe and Australia. Some of the early paints used natural earth pigments that are still in use today, including iron oxide, chalk, charcoal, and terra verde (Gooch 2002).

Beginning in the late 1700s, the Industrial Revolution resulted in the growth of the paint/coatings industry, since there were so many manufactured items produced during this period that needed to be coated. "However you look at it, paints and coatings have evolved from the simple ... colors on cave walls into a primary protective barrier between our possessions and our environment" (ACA 2014, p. 1).

A major breakthrough in household, do-it-yourself painting occurred in 1941 when the Sherwin-Williams Company developed the first commercially available interior, water-based paint. They named it Kem-Tone—"Kem" stands for "chemically evolved material." This paint was developed using casein, a milk protein. (For more information on casein, see Chapter 5, p. 31.) Kem-Tone became enormously popular because it was easy to apply, dried quickly, and was washable. More than 10 million gallons were sold in the first three years of production (ACS 2014).

During the last half of the 20th century, the paint industry, in response to government regulations, developed products that do not contain lead and have fewer volatile substances, making them more environmentally friendly. Lead-based paints were used for many years, in part because they are highly durable. However, the ingestion of lead-based paint by humans is harmful. Young children have been known to eat peeling paint and paint chips, which results in lead poisoning. The symptoms of lead poisoning in children often include developmental delay and learning difficulties. The ingestion of high levels of lead, in addition to causing brain damage, can affect the kidneys and nervous system. In the United States, the use of lead in household paints was phased out beginning in 1971 (and virtually banned in homes in 1978) as a result of federal legislation (Mayo Clinic 2014; see the "Resources" section [p. 61] for more information about U.S. standards for lead levels).

Investigating Coatings (Teacher Background Information)

Safety note: Students should wear an apron and indirectly vented chemical-splash goggles during the Explore phase of this lesson. Use natural water- or oil-based clay. Ensure proper ventilation when using all coating substances. Be sure that students do not mix the coatings, and follow school safety guidelines for the use, handling, storage, and disposal of hazardous chemicals. Also check for student allergies before assigning substances.

Materials

For the Engage phase of this lesson, you will need a small bag of six pennies for each group of four students. The pennies should include a range of shininess from bright to dull. Ideally, one will exhibit a tinge of green patina (see Figure 8.1). These pennies can be reused for each class. In addition, you will need photos of copper objects in different states of corrosion, including one of the Statue of Liberty.

During the Explore phase, students need to wear aprons and indirectly vented chemical splash goggles. Each group of students will need two 100 mm petri dishes (or margarine tubs with covers) and nontoxic modeling clay formed into four strips approximately $8 \times 1 \times 0.5$ cm (to support the coins in the petri dish). Each group will need about 20 ml each of tap water and 5% plain white vinegar, as well as 12 shiny pennies. The pennies in the Explore phase are not the same ones used in the Engage phase and cannot be reused from class to class. That is, each class will need its own set of bright, clean pennies. Have a variety of coatings available for

FIGURE 8.1   Assorted pennies

to produce the coins using zinc plated with copper, causing the percentage of copper in pennies to drop from 95% to just 2.5% (U.S. Mint 2014). You may wish to integrate social studies into this lesson by having students research why these changes were made. Students could also debate whether the penny should be discontinued, as has been done in Canada.

Note that the modern pennies will also undergo corrosion, since tarnishing and the formation of a patina are surface phenomena. The brownish color that comes about through a penny's exposure to the atmosphere is essentially caused by oxidation of the copper. In time, the copper oxide will further react to form a green patina, which may be evident in a few of your pennies (see Figure 8.2). The formation of the patina is hastened by an environment that includes acid rain, salt, or other pollutants. Interestingly, the patina on copper serves as a protectant from further corrosion. The patina is usually a mixture of several different

experimentation, for example, clear nail polish, car wax, paint, oils (cooking sprays or penetrating oils), polyurethane, petroleum jelly, water sealers, and lip gloss or balm. Students may request to use other types of substances for coating the pennies, but all materials should be approved by the teacher for safety purposes. Students will need newspaper to cover their work area as they apply their coatings. Cotton swabs can be used to cover one side of the pennies with the different coatings. Have small containers available to hold the coatings given to each group. Finally, students need to label how each penny is coated during the testing period. They can use a marker and masking tape or an index card for this purpose.

Engage

With the exception of pennies made in 1943 during World War II (when they were made of steel, as copper was rationed and used to produce shell casings for ammunition), copper has been part of the composition of pennies since 1793. In 1982, as the value of copper continued to increase, the United States began

FIGURE 8.2   Pennies showing (counterclockwise from top) little oxidation, oxidation, and patina

copper compounds, depending on the composition of the atmosphere.

Explore

Ask students what may be done to prevent copper from corroding. Students may initially suggest putting the pennies in a jar or another hermetically sealed container to prevent contact with the atmosphere. Other suggestions may include covering the pennies with plastic wrap or bags. The Explore question helps direct student focus to the use of coatings to protect the pennies.

This investigation will require two days. Although some corrosion will be observable in the acidic environment in about an hour, it is much easier to see after waiting overnight. Thus, plan on storing the petri dishes where they will not be disturbed for a day. Depending on the experience level of your students, you may wish to have them design their own experimental procedures. We have provided one example for students to follow (see Activity Worksheet 8.1, pp. 62–63). This may be a good time to review the key components of an experiment, such as identifying the independent and dependent variables and quantitative versus qualitative data. In the experiment described here, there are two independent variables: the type of atmospheric environment and the type of coating. The dependent variable is the change in the color and sheen of the coins. Students will be making qualitative observations.

Explain

The results will depend on the coatings used, but students will likely find that the acidic environment produces more corrosion on the pennies than the nonacidic one. Typical results for the coated pennies in the moist environment are shown in Figure 8.3 and in the moist, acidic environment in Figure 8.4.

Discuss with students what constitutes evidence that a chemical reaction has taken place. Common evidence includes color change, temperature change, the formation of a gas or a precipitate, odor, or decom-

FIGURE 8.3 Experimental results for coatings in a moist environment: (A) Petroleum jelly, (B) Household oil, (C) Lip balm, (D) Correction fluid, (E) Cooking spray, and (F) Control

position. In this case, there is an obvious change in color from the shiny copper to the shades of brown and the green patina. Students will note that some of the coated pennies did not experience the chemical change of corrosion.

Some students may observe that painting a penny represents a color change and may therefore incorrectly infer that this is evidence of a chemical change. Help these students understand that painting the coin is actually a physical change and to think of the paint as a physical barrier. One might try using an analogy of a more visible physical barrier, such as wrapping the coin in plastic wrap to prevent the corrosive reaction.

To help students reach an overall conclusion regarding the effectiveness of the various coatings, discuss all findings with the entire class. Ask students to analyze each of their coatings to assess benefits and

FIGURE 8.4

Experimental results for coatings in a moist, acidic environment: (A) Petroleum jelly, (B) Household oil, (C) Lip balm, (D) Correction fluid, (E) Cooking spray, and (F) Control

weaknesses. Engineers always need to consider the costs and benefits of their design solutions. For example, a good coat of paint may prevent a penny from corroding for a while. In that sense, it is an acceptable solution if one wishes to protect the copper. However, like anything that is painted, the paint will chip, fade, crack, and rub off and need to be reapplied, adding an ongoing expense. To determine the best solution, students will need to balance the various coatings' benefits and weaknesses and will likely come to different conclusions that will, ideally, lead to lively debate.

Extend

The purpose of this stage is to help students recognize the many types and uses of coatings in everyday life (see Table 8.1, p. 60). Some coatings are used to prevent corrosion, as they have just experienced. Other uses include promoting cleanliness or providing protection from physical damage, ultraviolet light, or moisture. Ask students to think of other everyday coatings; list the coatings' purposes, benefits, and weaknesses in a table; and discuss conclusions.

After students have discussed their tables, you may wish to extend the idea of costs and benefits in an economic sense. For example, furniture polish used on a wooden table may be considered expensive; however, the polish cost may become insignificant when compared to the value of a fine wooden table.

Evaluate

Ask students to apply their new knowledge about corrosion to the problem of preventing a steel bicycle chain from rusting. In their written responses, students should discuss the benefits and drawbacks of using various methods to stop rust from forming. Students will likely come up with a number of ways of protecting the chain, most of which probably involve using some sort of oil. In general, the benefits of oil are that it not only protects the chain from rust but also lubricates it so that the individual links move easily. Weaknesses are that oil is messy, an excess of oil can collect dirt, and the oil must be reapplied regularly. Alternatively, students may suggest covering the bicycle in some way—perhaps with a tarp or some other material. The advantage of covering the bicycle is that it will keep rain, snow, and dirt off of the bike; the disadvantage is that the cover will not seal the bicycle completely from moisture and the chain will likely still rust, albeit more slowly.

Conclusion

The use of coatings has a history dating back more than 30,000 years, with paint in particular becoming a major industry in the last 300 years or so. Coatings have evolved because of continual engineering to design materials to meet the constraints of different coating problems. Indeed, coatings are very much a part of

TABLE 8.1   Sample coatings

Item	Protective coating	Purpose	Benefits	Weaknesses
Wood floor	Polyurethane, varnish, wax, etc.	Scratch and moisture resistance; sheen	Durable	Will scratch off in time
Wooden table	Tablecloth or placemat	Scratch and moisture resistance	Easily removed or replaced	Must be on table each time it is used
Automobile	Paint	Corrosion prevention, appearance	Effective	Easily scratched
Automobile	Wax	Paint protection, sheen	Keeps paint shiny	Must be reapplied often
Outdoor metal furniture	Powder coating	Corrosion resistance, appearance	Applied with no volatile liquids	Will scratch like paint
Skin	Sunscreen	UV protection	Protects skin from damage	Wears off, especially in water
Interior walls	Paint	Appearance, ease of cleaning	Washable, attractive	Can become soiled or damaged by water
Aluminum cookware	Thin oxide layer (anodization)	Appearance, protection	Appearance, durability	Must be hand washed with care

our everyday lives, ranging from high-tech paints to polishes and waxes to lip balms to a simple cap to cover and protect your head.

References

American Chemical Society (ACS). 2014. Waterborne interior paint: Kem-Tone wall finish. National historic chemical landmark. *www.acs.org/content/acs/en/education/whatischemistry/landmarks/kem-tone.html*

American Coatings Association (ACA). 2014. History of paint. *www.paint.org/about-our-industry/history-of-paint*

Copper Development Association (CDA). 2014. Reclothing the first lady of metals—Repair concerns. *www.copper.org/education/liberty/liberty_reclothed1.html*

Gooch, J. W. 2002. *Lead-based paint handbook.* New York: Springer.

Mayo Clinic. 2014. Diseases and conditions: Lead poisoning. *www.mayoclinic.org/diseases-conditions/lead-poisoning/basics/symptoms/con-20035487*

NGSS Lead States. 2013. *Next Generation Science Standards: For states, by states.* Washington, DC: National Academies Press. *www.nextgenscience.org/next-generation-science-standards*

U.S. Mint. 2014. The composition of the cent. *www.usmint.gov/about_the_mint/fun_facts/?action=fun_facts2*

Resources

Lead toxicity: What are the U.S. standards for lead levels? *www.atsdr.cdc.gov/csem/csem.asp?csem=7&po=8*

The Statue of Liberty documentary (video). *https://www.youtube.com/watch?v=j6NSq3t_EHY*

Statue of Liberty (History Channel video). *www.history.com/topics/statue-of-liberty*

CHAPTER 8

ACTIVITY WORKSHEET 8.1 — Investigating Coatings

Engage

1. Observe the pennies your teacher has provided. How do the pennies differ from one another? How do the color and sheen of the pennies vary? Record your observations.

2. Look at the pictures your teacher shows you of other items made of copper. How do these compare to your copper pennies?

3. What are some possible reasons for the variations in the sheen of the copper items?

4. If you wanted to keep the copper pennies shiny, what might you do? Record your ideas. In this exploration, we will focus on investigating the question of how a penny might be coated to prevent it from corroding.

Explore

Safety note: Wear indirectly vented chemical splash goggles and an apron throughout this exploration. Do not mix any of the coatings together. Be sure to follow your teacher's instructions when disposing of the substances at the end of the exploration.

1. You will set up a testing procedure to evaluate different coatings that may protect the pennies from corroding. Of the coatings available, select the five you wish to test.

2. Coat the top surface of two pennies with one of the five coatings and set them aside to dry. Repeat this process four times using each of the other coatings. Be sure to label all the pennies so you know which of the coatings is on each. You will use the two remaining pennies as controls to see what happens without any coating.

3. Using the materials provided, you should set up two containers in which to test the coatings. Make four strips of clay about 8 × 1 × 0.5 cm. Place two parallel strips in each dish approximately 3 cm apart. The strips are intended to keep the pennies from direct contact with the liquid that will be added. Place three pennies on each of the clay strips, and label the type of coating on each penny. Make sure that there is one penny of each coating type and one control penny in each dish.

4. For the moist environment, one dish will contain 20 ml of water; the other will contain 20 ml of ordinary white vinegar for the moist, acidic environment. Pour the liquids into the two dishes, using care not to splash the top surfaces of the pennies. Cover the two dishes, making sure the lids do not touch the pennies. Set the dishes aside until the next day.

Explain

1. Examine your coins. Note if the sheen or color of each has changed overnight. Record your observations in a table like the one shown below.

Coating	Moist environment	Moist, acidic environment
Control		
Coating 1		
Coating 2		
Coating 3		
Coating 4		
Coating 5		

2. How do the pennies in the moist environment compare to the pennies in the moist, acidic envi-

ronment? What evidence do you see that would indicate a chemical reaction has taken place?

3. Which coatings resulted in the least amount of corrosion? The most?

4. What is the condition of the uncoated control penny in each environment?

5. Share your data and compare your results with your entire class.

6. Some oils work well to prevent corrosion. However, you might not want to carry oily coins in your pocket. The benefit of this coating is that the oil may prevent corrosion; the weakness is that it is messy. Evaluate each of your coatings for benefits and weaknesses.

7. Based on your analysis, which coating is best? In engineering terms, this is considered the *optimal design*.

Extend

1. In this investigation, we have considered the use of coatings to prevent the corrosion of copper. Many objects in our everyday lives make use of a coating that is designed to provide some type of protection. For example, you might coat a wooden table with furniture polish. Make a data table with five columns like the one in the table below. In the first column, list at least five everyday items that are coated, and in the second column, list the coating that provides some type of protection to that item.

2. Discuss with your group the purpose of each coating and list these in the third column. For example, the furniture polish may help protect the wooden table from wet cups and glasses. For each item on your list, record the benefits and weaknesses of each coating in columns four and five.

Evaluate

Rust is a form of corrosion that occurs to items made of steel. Suppose someone's bicycle chain is getting a bit rusty because there is no place to keep it indoors. What are at least two ways you would suggest to protect the bicycle chain from rusting? Write an explanation of the benefits and weaknesses of each method.

Item	Protective coating	Purpose	Benefits	Weaknesses
Wooden table	Furniture polish	Protect from moisture	Keeps wood from being stained	Must be applied frequently

PART 3

Engineering at the Retail Store

CHAPTER 9

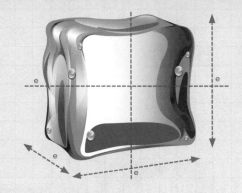

SHOULD ICE BE CUBED?

WHAT KIND OF ice do you put in your summer drink? Do you prefer crushed, shaved, or cubed? Most in-home, automated ice makers produce crescent-shaped ice. For some, ice is ice, but for others, the shape matters. If you have a preference, is it based on aesthetics or physics? Perhaps you prefer the look of one cube over another in terms of how they stack in a glass. Maybe you enjoy novelty-shaped cubes (see Figure 9.1). Others may choose one shape over another for scientific reasons—do you want ice that cools your drink the quickest or dilutes it the least?

While ice is usually referred to as ice cubes, indeed, most are not really cubes at all. In this 5E learning-cycle lesson, students will investigate different shapes of ice and how shape affects the speed of melting and the rate of cooling a glass of water. Students will compare three different shapes of ice with the same volume (10 cm³) but different surface areas. The *Next Generation Science Standards* (*NGSS*) state that middle-level students should learn that "modeling, testing, evaluating, and modifying are used to transform ideas into practical solutions" (ITEEA 2007, p. 103). This standard introduces the engineering practices of testing and evaluating while students investigate ice melting rates and ice tray design. As stated in the *NGSS*, students should understand that "[t]he changes of state that occur with variations in temperature or pressure can be described and predicted using these models of matter" (MS-PS1-4; NGSS Lead States 2013). Students will discover that melting is a phenomenon that takes place at the surface of a solid and that melting rate is proportional to surface

FIGURE 9.1	Novelty ice cube trays

area. This activity can be integrated with a unit dealing with states of matter.

Historical Information

People have used ice to preserve food for many centuries. Beginning around 1700, ice was harvested from lakes and ponds in the winter, then stored in sheds called icehouses for use in the summer. By the 1800s, iceboxes were introduced for home use. Blocks of ice were placed near the top of the box, where the cooler, denser air circulated through it (Fairbanks Museum & Planetarium 2003). The use of iceboxes continued until refrigerators became common in the 1930s. In 1932, Guy Tinkham patented the first flexible ice tray, which allowed for the

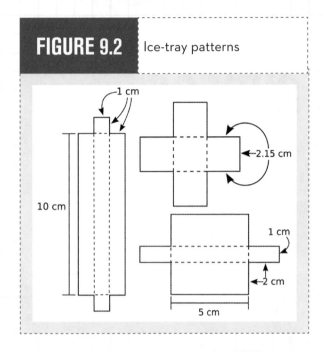

FIGURE 9.2 Ice-tray patterns

easier removal of ice (Marcus 2007). At about the same time, inventor Lloyd Copeman was walking through slush and snow and noticed that the ice could be easily knocked off of his rubber boots. He used this experience to design ice cube trays made from rubber so the cubes could be readily removed (Clever 1954).

The use of the term *ice cube* can be traced to Copeman, and while he may not have been the first to use it, the term persists to this day. Ice cube trays are no longer as important as they once were. Automatic ice cube makers have been installed in household refrigerators since the 1950s and have become more and more popular over time—at least for Americans, as Europeans do not usually put ice in their beverages.

Investigating Ice Cube Shape (Teacher Background Information)

Materials

In the Engage phase, each group of students will need a ruler and an ice cube tray to determine the shape (and possibly the volume) of a typical ice cube. Ice cube trays are available at dollar stores and can be reused with each class. The Explore activity takes place over two days—the first day will be for constructing the ice cube trays and the second for testing the ice that the trays produce. Students should work in groups of three, if possible, on the second day. To construct the ice cube trays, each group will need transparent plastic report covers, scissors, a ruler, and about 1.5 m of duct tape, as well as a fine-tipped waterproof marker to scribe the plastic and the pattern for the three trays (see Figure 9.2). You will need to replenish the plastic report covers and duct tape for each class. You will also need a cookie sheet or similar container to hold the trays in the freezer overnight.

On the second day, for testing, each group will need three identical, clear containers that have an opening greater than 10 cm to accommodate the longest ice cube. Plastic disposable food containers that can hold about 400 ml of water work well and can be used for multiple classes. Students will also need a stopwatch and chemical splash goggles.

For the Extend phase, each group of students will need two of the food containers filled with 400 ml of room-temperature water, two digital thermometers (if possible) to allow for precise temperature readings, and two stirring rods or plastic spoons. Each group will also need two small plastic bags containing eight ice cubes each—one bag of crushed cubes and one of whole cubes. Use a small hammer or mallet to crush the ice ahead of time. Put the ice to be crushed into a double baggie, as the bags tend to rip apart when hit with the hammer.

Engage

Initiate a discussion with students regarding ice cubes they use at home, directing them to think about the size and shape of the ice cubes. It is likely that they will realize that most household ice cubes are not really cubic in shape. To confirm, distribute a ruler and ice cube tray to each group and have them measure the height, width, and breadth of one opening. Most are rectangular prisms but rarely cubic (see Figure 9.3). If you do not have sufficient trays, then distribute ice cubes to each group for measuring.

The idea for this column originated when one of the authors was in a specialty kitchen store and noticed a display of ice cube trays designed to make extra-large cubes. The larger cubes will melt more slowly than an equal volume of smaller cubes because there is less surface area on one large cube. The result is an ice cube that will not water down a beverage—although, as we will see, there is a trade-off: The larger cube will take a bit longer to cool the beverage. What struck the author, however, was the advertising copy on the package, which said that the "cubes' large surface space is designed to melt slower than standard cubes." Clearly, the manufacturer meant to say large volume and small surface area would cause the cube to melt more slowly. Many students share this same confusion with the ideas of volume and surface area and therefore require multiple experiences with this topic (e.g., see Moyer and Everett 2010). At this point, you may need to clarify the difference between surface area and volume for your students. In this exploration, students will hold volume constant and vary the surface area.

Do not share this answer with students at this time, since their engineering challenge is to test and evaluate the effects of different-shaped pieces of ice with different surface areas but the same volume (10 cm^3).

Explore

The patterns for three different-shaped trays are provided in Figure 9.2. (Full-size patterns are available online at *www.nsta.org/more-engineering*.) You may give students copies of these patterns or provide the dimensions and devote class time for students to make their own. Be sure students realize that the three different shapes have the same volume: 10 cm^3. Students need to use sufficient duct tape to ensure that their trays do not leak. You may wish to collect all the trays in a large baking pan or similar container, fill them at the end of the day, and place them in a freezer. If you want to

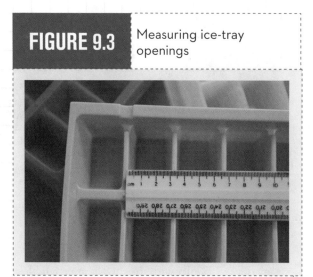

FIGURE 9.3 Measuring ice-tray openings

ensure that each group receives its own trays back for testing, have groups label their trays with an indelible marker. To maintain the two rectangular shapes as the ice expands upon freezing, you need to use a piece of tape across the top, as shown in Figure 9.4. At this point, direct the students to calculate the volume and the surface area for each shape. Students typically can calculate the volume of a rectangular prism ($L \times W \times H$)

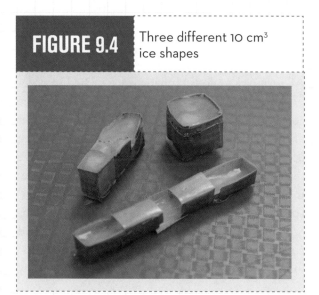

FIGURE 9.4 Three different 10 cm^3 ice shapes

Dimensions of tray (cm) L x W x H	Volume (cm³)	Surface area (cm²)	Time to melt (seconds)
10 x 1 x 1	10	42	135
5 x 1 x 2	10	34	215
2.15 x 2.15 x 2.15	10	27.7	325

TABLE 9.1 Sample data

but may have difficulty with the surface area, which is the sum of all six surfaces and is found by the following:

$$A = 2(L \times W) + 2(L \times H) + 2(W \times H)$$

On day 2, students should set up their testing materials prior to getting the ice from the freezer. The three containers should each contain 400 ml of room-temperature water. To ensure the water is the same temperature, have each group fill a pitcher and then measure 400 ml into each container. Students can then place one piece of ice in each container and measure the time it takes for the ice to melt. Sample data are shown in Table 9.1.

Explain

All groups should find that the ice melted in order of decreasing surface area. That is, the greater the surface area, the faster the ice melts. The transfer of thermal energy from the warmer water to the ice occurs from outside of the ice inward. Therefore, since the cube had the smallest surface area of the three, it takes the longest time to melt. In fact, a 10 cm³ spherical piece of ice would take even longer to melt. Students should conclude that the volume was held constant in the testing to isolate surface area as the independent variable. Ask students to speculate why

ice cube trays have different shapes; they may suggest numerous reasons.

Extend

Again, students need to set up their materials prior to getting their ice and also make sure they begin with the same volume of approximately room-temperature water. Distribute the bags of crushed and cubed ice and alert students to be ready to make careful observations as soon as the ice is added. As shown in Figure 9.5, after 30 seconds, the crushed-ice mixture had dropped 9 degrees from the starting temperature of 22.4°C, while the temperature of the water with the cubes' only dropped slightly more than 1 degree. Figure 9.6, taken after three minutes, shows that the crushed ice is nearly melted and the temperature of the water with the cubes is beginning to catch up. They will eventually reach the same temperature. Students should conclude that engineers may purposely design small cubes of ice if the outcome desired is rapid cooling. The cost of this benefit, however, is that the smaller pieces of ice melt much more quickly and will therefore dilute the beverage. Conversely, large pieces of ice will cool at a slower rate but will last longer and, initially at least, dilute less. This Extend activity can be conducted as a demonstration if obtaining enough materials is problematic.

FIGURE 9.5 Crushed versus cubes to measure cooling rate (30 seconds)

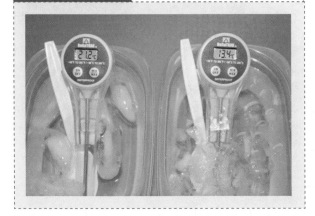

FIGURE 9.6 Crushed versus cubes to measure cooling rate (3 minutes)

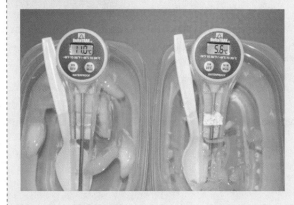

Evaluate

Students should now realize that the phrase "large surface space" on the label the author saw is actually misused. The term *surface* refers to an area, whereas *space* implies some volume. (When students rewrite the advertising copy, as instructed in Activity Worksheet 9.1 [pp. 73–74], pay close attention to their choice of words as they distinguish between *surface area* and *volume*.) Students should now understand that a piece of ice with a smaller surface area will melt more slowly. For any amount of ice, if it is amassed into one piece and shaped into a cube, it will then melt at the slowest rate (see Figure 9.7). For this reason, when people used iceboxes to refrigerate their food, large blocks of ice were used. Ten pounds of ice cubes will melt faster, for example, than a 10-pound block of ice.

Conclusion

All four STEM areas were integrated in this activity. Science was represented through heat transfer, cooling rates, and melting. Technology was integrated through the use of the different types of ice cube trays—novelty, giant cube, household, and student built. The activity allowed students to become engineers by testing three

FIGURE 9.7 Extra-large ice cube versus standard ice cubes

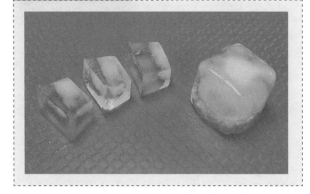

different ice-tray designs. Finally, the students used mathematics when they measured the surface-area and volume relationships among the ice cubes. The integration of STEM areas will help move your teaching toward the recommendations of *A Framework for K–12 Science Education: Practices, Crosscutting Concepts, and Core Ideas* (NRC 2012). The wide variety of ice cube trays available is not really just random design, but is actually something that is engineered to meet specific needs.

References

Clever, A. K. 1954. Fifty years an inventor. *Popular Mechanics* 102 (5): 108–10.

Fairbanks Museum & Planetarium. 2003. *Ice harvesting*. St. Johnsbury, VT: Fairbanks Museum & Planetarium.

International Technology and Engineering Educators Association (ITEEA). 2007. *Standards for technological literacy: Content for the study of technology*. Reston, VA: ITEEA.

Marcus, E. *Baltimore Sun*. 2007. The clear facts about ice cubes. March 14.

Moyer, R., and S. Everett. 2010. Everyday engineering: What makes a better box? *Science Scope* 33 (6): 64–69.

National Research Council (NRC). 2012. *A framework for K–12 science education: Practices, crosscutting concepts, and core ideas*. Washington, DC: National Academies Press.

NGSS Lead States. 2013. *Next Generation Science Standards: For states, by states*. Washington, DC: National Academies Press. *www.nextgenscience.org/next-generation-science-standards*

ACTIVITY WORKSHEET 9.1 Investigating Ice Cube Shapes

In this activity, you will investigate whether surface area affects the melting rate of ice cubes.

Engage

1. What shape are the ice cubes at your home? What shapes do others in the class have? Why do you think ice cubes come in different shapes? What are some possible effects based on the shape or size of an ice cube? Record your ideas.

2. Are all ice cubes actually cubic? Measure the opening for one ice cube in a typical ice cube tray. What are the dimensions for height, width, and breadth? Will this tray produce a cubic piece of ice? Explain or give reasons for your answer. Record your responses and share with your classmates.

3. In a store, the package on an ice cube tray that makes very large ice cubes included the following statement: "The cubes' large surface space is designed to melt slower than standard cubes." What do you think the manufacturer of the trays means by this statement? Record your thoughts and then discuss with your group.

4. How might you test the melting rate of different-shaped ice cubes?

Explore

1. Your teacher will provide you with the patterns for three different shapes of ice cube trays.

2. Using the materials, build the three different ice cube trays. Carefully tape each tray so it does not leak. Fill each with 10 ml of water and place in a freezer overnight.

3. Complete the table on the next page. You will need to calculate the surface area and volume of each tray.

4. Do you think all of the ice cubes will melt at the same time, or will their melting rates be different? Include your reason for making this prediction, and record your prediction.

5. To conduct the test, you must prepare three containers of water before removing the ice from the freezer or the trays. Obtain three containers with 400 ml of room-temperature water. Why is it important that the temperature of each container is the same?

6. At the same time, carefully remove the three pieces of ice from the trays and place each in a separate container of water and start timing. Time how long it takes for each ice cube shape to melt. You will need to observe carefully to determine when they have completely melted. Record the times in your data table.

Explain

1. In what order did your cubes melt? How does this compare with the results of other groups?

2. What do you notice about surface area and melting time? Explain.

3. Why do you think the volume was held constant in this experiment?

4. What might be some reasons ice cube trays are designed to have different shapes?

Extend

1. Some people may prefer crushed ice rather than ice cubes. Why might someone want to use crushed rather than cubed ice? Will crushed and cubed ice melt at the same rate? Will one cool a glass of water more quickly than the other? Will one keep a drink cold longer? Discuss with your group and then record your predictions.

2. Obtain two containers and fill each with 400 ml of room-temperature water. Place a thermometer in each container, and check and record the temperature of each.

3. Your teacher will provide you with samples of crushed and cubed ice. At the same time, place one sample in each container, stir once, and start timing.

4. Measure and record the temperature every 30 seconds for five minutes, stirring once before taking each temperature reading.

5. How do your results compare to your predictions?

6. Under what conditions might an engineer design ice trays that make very small pieces of ice? Large pieces of ice?

Evaluate

Consider again the advertising on the package of the supersized ice cube trays we discussed in the Engage section. What do you think the manufacturer meant to say instead of "large surface space"? Using what you have learned from the testing you have done, create some new advertising for the package that is more scientifically correct.

Dimensions of tray (cm) L x W x H	Volume (cm³)	Surface area (cm²)	Time to melt (seconds)
10 x 1 x 1			
5 x 1 x 2			
2.15 x 2.15 x 2.15			

CHAPTER 10

IT'S STUCK ON YOU

AT THE DRUGSTORE, what goes through your mind as you select a box of adhesive bandages? The choices are numerous. Do you prefer plastic or fabric? Medicated or plain? Waterproof, sheer, or cartoon laden? What about the size or shape—strips, squares, or small round spots? Are you interested in latex-free bandages or perhaps special adhesives for sensitive skin? There are dozens of choices, all designed for some specific purpose. Once again, we find a rather simple everyday product that has been engineered to meet different everyday human requirements.

In this 5E learning-cycle lesson, students will test adhesive bandages and then design a bandage for the specific function of keeping a wound on their knuckle dry. The *Next Generation Science Standards* note, "A solution needs to be tested, and then modified on the basis of the test results, in order to improve it" (MS-ETS1-4; NGSS Lead States 2013). The standards state that the practice of design and testing in engineering is similar to the practice of scientific inquiry. Both require planning—problem solutions in engineering and investigations in science, data collection, and recording results (MS-ETS1-1).

Historical Information

The first record of humans caring for wounds can be found in writings from Egypt, around 2000 BCE. Wound dressings at that time consisted of lint, grease, and honey. Lint and grease were used as a covering, and the honey was used to reduce infection (Forrest

1982). Sterile surgical bandages date back to the late 1800s, but the invention of the prepackaged bandage (e.g., Band-Aid) did not occur until 1920. Before this, people made a small bandage of cotton gauze and adhesive tape to cover small cuts. Earle Dickson, an employee of Johnson & Johnson (a producer of medical supplies), is credited for the invention of premade bandages. His wife was said to be accident prone. To facilitate her tending to her own cuts and scrapes while he was working, he took a long piece of adhesive tape and placed swatches of the cotton gauze at intervals along it and then covered the entire strip with crinoline so it would not stick to itself. All she had to do was cut a piece off, remove the crinoline, and apply the bandage to her wound. His employer decided to market Dickson's idea, and the first Band-Aids came to be, not in the strips with which we are familiar, but in a roll similar to cellophane tape. Like many new ideas, it took a while for people to recognize their convenience. They did not sell well when they first became available, so the company actually gave them away to scouting groups and the military (Johnson & Johnson 2012).

Investigating Bandages (Teacher Background Information)

Materials

In the Engage phase, each group of three to four students will need four different types of bandages. Be sure to include at least one fabric bandage, one plastic strip, and one that is labeled waterproof. You might also include

CHAPTER 10

some name brands and some generic or store brands. These can be purchased locally for about $2 to $3 for a box, depending on type, of 20 to 60 bandages. In the Explore phase, each group will need at least four bandages—two identical ones that are waterproof and two identical ones that are not waterproof. They will also need a plastic bowl (about a liter in capacity) about two-thirds full of room-temperature coffee. The exact amount is not critical but should allow a student's hand to be submerged without spilling. You can use colored water if you desire, but food dyes are prone to stain hands and clothing. Paper towels should also be available for drying hands after testing. You may also wish to have newspapers available to place on tables to absorb spills or drips. In the Extend phase, groups will need general school supplies for planning their designs, scissors, at least one 4 × 4 in. gauze pad, about 80 cm of 2 in. (5 cm) wide waterproof adhesive tape (cloth medical tape and athletic tape are not waterproof), and a sheet of waxed paper. The gauze pads are about $8 for a box of 100, and the 2 in. (5 cm) wide tape is about $3 for 5 yards (4.5 m). For the Evaluate phase, each group will need one knuckle bandage, which costs about 15 cents each. *Safety note:* Many bandages are made with latex products. Be sure to check for student allergies. Students should wear chemical splash goggles during this activity.

Engage

Show students a variety of different types of bandages, and elicit their prior knowledge and experiences with the various types. Discuss how they think the properties of bandages relate to how well they function. For example, cloth bandages have the advantage of comfort and flexibility but the weakness of easily becoming water soaked and frayed. Once again, we see, in engineering terms, that most products that have been designed have both costs and benefits. Distribute three or four different types of bandages to each group, and have students look for similarities and differences (do not distribute knuckle bandages at this point, as they are the focus of the Evaluate phase of this lesson).

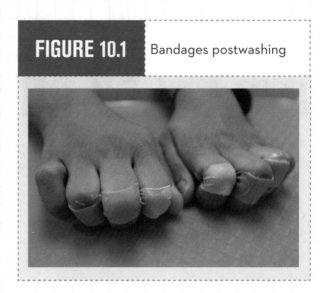

FIGURE 10.1 Bandages postwashing

Students should notice that most are of the same rectangular shape with a gauze pad. There are, of course, many bandages of specialized shapes—for fingertips, knuckles, and larger or smaller wounds. Students may notice that in some bandages (usually the waterproof ones), the adhesive surrounds the gauze pad on all sides, whereas others have no adhesive on the two sides parallel to the bandage's length.

The hand-washing test will help students identify some of the strengths and weaknesses of the bandages related to durability and adhesive strength. Students will notice that most bandages will pucker on areas of the hand that flex, most noticeably on the knuckle (Figure 10.1). Most cloth bandages we have tested fray on one hand-washing and absorb water and become rather soggy. The adhesive on different brands and types varies such that some bandages can actually slip off of a student's finger when wet.

Point out to students that some manufacturers claim on their packaging that the bandages are waterproof. In this exploration, students will design a test to determine whether a bandage is waterproof. The method outlined in Activity Worksheet 10.1 (pp. 80–81) is patterned after one developed by *Consumer Reports* (2010).

FIGURE 10.2 Waterproof testing in coffee

FIGURE 10.3 Testing results—dry and wet

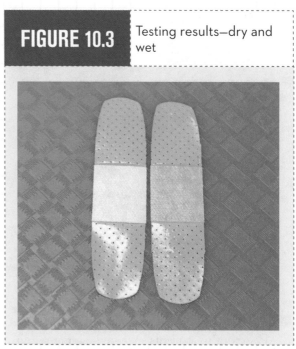

Explore

In the Activity Worksheet, we have provided the outline of the *Consumer Reports* testing but left some of the experimental design for students to plan. You may wish to provide more scaffolding if your students are not accustomed to designing their own experiments. In that case, we recommend placing each of the bandages on two different positions on the hand—perhaps the back of the hand, which is relatively flat, and a second one on a knuckle. We have found that submerging the hand for about 30 seconds produces good results (Figure 10.2). We had students flex their hands while submerged. Students should dry their hands after removing them from the coffee. Take off the bandages and inspect the gauze portion for brown tinting, which is an indication of moisture penetration and lack of waterproofness (Figure 10.3).

Explain

Typical results will show that nonwaterproof bandages have completely soaked gauzes regardless of where on the hand they were located. Waterproof bandages usually show mixed results. For example, they often keep most water out when placed on the flatter surface of the back of one's hand. When used on a knuckle, however, our results indicate that most will leak to some extent. This finding is also supported by *Consumer Reports* (2010). Students should infer that the only waterproof bandages are those with adhesive that completely surrounds the gauze. In addition, they should realize that none of the fabric bandages are waterproof.

The glue on a bandage sticks because of a property known as adhesion. *Adhesion* is when something sticks to other objects; *cohesion* is the sticking of similar things to one another. Bandages exhibit both properties. Adhesion takes place at the molecular level, primarily as a result of electrostatic attraction. To make a bandage waterproof, the glue, or adhesive part, must completely surround the gauze, totally bonding with the skin, to seal out the moisture. Unfortunately, the glue also adheres to any hair follicles, which adds to the pain when you remove the bandage.

FIGURE 10.4 Different student-designed knuckle bandages

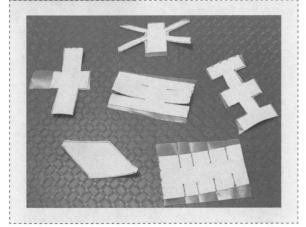

FIGURE 10.5 Bandages designed for two knuckles

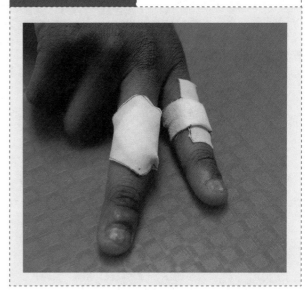

Point out to students that engineering requires constant retesting based on what is learned. Sometimes you might modify your testing procedure, and other times you might modify your design. In this lesson, students are asked to do both. In the Explain phase, students are asked to critique the experimental procedure, and in the Extend phase, they are asked to modify the design of the bandages.

Extend

In this phase of the lesson, students will use what they have learned in previous phases to design a bandage that will keep their knuckle dry. To encourage design creativity, do not show students a commercial knuckle bandage at this time. Have students draw their designs on paper and include their thinking as to why it should work.

Have students put the adhesive tape on a piece of waxed paper, transfer their design to the back, and then carefully cut it out (Figure 10.4). Do not remove the waxed paper until you are ready to put on a piece of gauze and prepare for testing with the coffee (Figure 10.5). Students should realize that no design is perfect as they share their results. Typically, students

will find that most designs give up finger flexibility for increased waterproof ability.

If you want to extend the lesson into the STEM area of mathematics, you might have students devise a method of determining the strength of a bandage and then look at the ratio of strength to the unit cost of each different bandage. The higher the ratio, the "better" the bandage.

Evaluate

Finally, show students a manufactured knuckle bandage. These are usually made with a pad with four tails that go under the finger, leaving the bottom of the knuckle able to flex. They are designed this way to allow greater flexibility of the knuckle. The gauze pad is completely surrounded by adhesive. Have students plan a way to compare their design to the manufactured one. They will likely conduct a coffee test much like the one they did previously.

Conclusion

In this lesson, students learn about bandage design and testing, as well as some costs and benefits associated with a design. Of course, not all students will become engineers, but engineering practices also have the benefit of providing students the opportunity to become critical consumers—a lifelong *everyday* skill.

Acknowledgment
The authors wish to thank their colleague LaShorage Shaffer.

References

Consumer Reports. 2010. Few waterproof bandages make the cut. November. *www.consumerreports.org/cro/magazine-archive/2010/november/health/waterproof-bandages/overview/index.htm*

Forrest, R. D. 1982. Early history of wound treatment. *Journal of the Royal Society of Medicine* 75 (3): 198–205.

Johnson & Johnson. 2012. Band-Aid brand heritage: Band-Aid brand adhesive bandages beginnings. *www.band-aid.com/brand-heritage*

NGSS Lead States. 2013. *Next Generation Science Standards: For states, by states*. Washington, DC: National Academies Press. *www.nextgenscience.org/next-generation-science-standards*

Resources

Kilmer House: The story behind Johnson & Johnson and its people. *www.kilmerhouse.com/category/did-you-know*

Science Scope April/May Integrated Instructional ... playlist. Playlist of related videos compiled by the editor. *www.youtube.com/playlist?list=PL0A051E56FD37D618&feature=mh_lolz*

ACTIVITY WORKSHEET 10.1   Investigating Bandages

Engage

1. At some point, everyone needs to use a Band-Aid. What are some properties that you look for in a good bandage? Discuss your thinking with the class. What are some problems you have experienced with adhesive bandages?

2. Your teacher will provide different types or brands of adhesive bandages. Record how these bandages are alike and how they are different.

3. Have one member of your group put a different bandage on the knuckle of several different fingers. One test of a bandage is how well it sticks after getting wet. Students with bandages should wash their hands for at least 20 seconds using a mild soap. (Students washing their hands need to wear chemical splash goggles because this is part of the lab procedure.) After drying, inspect each bandage's condition and record your observations. Discuss with your group how well each bandage held up.

4. Some bandage manufacturers claim that their bandages are waterproof. Discuss with your group the following explorable question: How might you test to see whether a bandage is indeed waterproof? Note that a bandage is either waterproof or not. If any liquid gets onto the gauze pad, then it is not waterproof.

5. One test was conducted by *Consumer Reports* (2010), which had people put their bandaged hand in room-temperature coffee. When they removed their hand and took off the bandage, they could tell whether it leaked if the gauze was tinted brown by the coffee. So, if there is brown tint on the gauze, the bandage has leaked and is not waterproof. Students must wear chemical-splash goggles when testing the coffee.

Explore

1. Design a science investigation to determine whether different bandages are waterproof. Test several different types of bandages, each one at two places on your hand. Think about the following questions as you design your experiment: How long will you submerge the bandaged hand in the coffee? Will you move your hand while it is in the coffee? How will you be sure not to get the gauze wet when you remove the bandage from your hand?

2. After your teacher approves your plan, carry out your investigation and record your data in a table.

3. Use the evidence to draw some conclusions to answer the following explorable question: Which bandages are waterproof?

Explain

1. Discuss your conclusions with the class. Make a list of the different findings of the entire class and the evidence to support them. What conclusions can you draw about which bandages are waterproof now that you have heard about your classmates' experiments? How might you change your experiment based on what you have heard about the others?

2. What are the characteristics of a waterproof bandage?

3. Reflect on the procedures you used in your experiment. What weaknesses were there in your experimental design? How might you do the experiment differently a second time?

Extend

1. Now that you have discovered some of the issues associated with waterproof bandages, your engineering challenge is to design a waterproof

bandage for use on the middle knuckle of your index finger.

2. Considering the materials provided by your teacher, make a written plan and a sketch of your design for a waterproof bandage to be used on your knuckle.

3. After your teacher approves your plan, construct and test your design. After initial testing, you may wish to modify your design and retest.

4. Share your design and results with your classmates. What types of designs worked best on a knuckle? What are the weaknesses of some of the designs? Are any of the designs free of weaknesses?

Evaluate

Imagine you decided to manufacture your knuckle bandage. Therefore, you need to test to see how your design compares with currently available products. Your teacher will provide your group with a waterproof bandage that has been designed for use on knuckles. Plan a way to compare this manufactured bandage with one that has already been designed by someone in your group. Test and present your findings and conclusions in writing.

Reference

Consumer Reports. 2010. Few waterproof bandages make the cut. November. *www.consumerreports. org/cro/magazine-archive/2010/november/health/ waterproof-bandages/overview/index.htm*

CHAPTER 11

QUEUING THEORY— IS MY LINE ALWAYS THE SLOWEST?

THE ANSWER TO the title's question is: probably. Why do you think this is the case? How do you choose your checkout line at the grocery store? Perhaps you are careful not to get behind the person with 15 items in the less-than-10-items line. Or, do you try to guess who will argue with the clerk about an item's price, requiring a manager to step in to adjudicate? Worse, might there be someone in your line who is actually going to write a check? It is likely that you have also experienced different kinds of lines, or *queues*. Your bank, for example, probably uses one long, serpentine line in which the first person goes to the next available teller. Airports and amusement parks usually use this system as well. Is there a difference in the efficiency of each model? The study of waiting in lines is part of a field of systems engineering called queuing theory.

In this 5E learning-cycle lesson, students conduct simulations to test several different models for processing customers through a checkout system. The *Next Generation Science Standards* state that students should be familiar with developing and using models and that at the middle level, students are to "[d]evelop and use a model to describe phenomena" (MS-PS4-2; NGSS Lead States 2013). This lesson shows students an iterative process that allows them to gather evidence to evaluate different models, as the standards suggest

that students at this level should "[e]valuate competing design solutions" (MS-ETS1-2).

Historical Information

While it is likely that people have been standing in various types of queues for centuries, the first recorded scientific study of waiting was conducted in Denmark in 1909 by the Danish engineer and mathematician Agner Krarup Erlang, who worked for the Copenhagen Telephone Company. He was trying to determine how many switches would be needed as the company introduced a more automated system to meet the increased number of customers. Erlang was searching for a system that would provide the shortest possible waiting time to connect customers to their calls. Erlang developed his theories from classic probability, and his work is still the basis for some of the fundamental mathematics of queuing theory today (Plus Math 1997).

Thousands of scholarly papers have been published on queuing theory since that time, but the basic problem discussed in all of them is similar: How many workers or stations are needed at, for example, a fast-food restaurant, given a certain number of expected customers, so that the time each customer must wait is perceived as an acceptable level? At the same time, the restaurant would be responsible for the additional cost of extra employees. For example, to ensure that 100% of the customers would not have to wait in line at all, the restaurant would need a number of employees equal to the expected number of customers at any one time. This would, of course, result in many employees on the payroll who are idle at various times—not a very efficient method.

On one hand, the question of queuing theory seems quite simple—customers expect some sort of service and often have to wait in line for the service (checking out, seeing the doctor, getting their car washed, or eating lunch). However, the process can be complex when we consider various compelling factors: people arriving at different times, the cost of losing a customer due to a long wait, the cost of providing the service, and so forth. Consequently, the mathematical models can be quite complicated.

Richard Larson is an engineer at the Massachusetts Institute of Technology (MIT) who studies the psychological aspects of queuing theory. He has found, for example, that people are willing to endure longer waiting times under certain conditions. He notes that in the 1950s, when more and more people were living and working in high-rise buildings in larger cities, many people complained about elevator queue times during rush hours. One experiment showed that adding full-length mirrors in elevator waiting areas resulted in a drop in the number of complaints to nearly zero. People had something to do to fill their time. He also speaks of the positive psychological effects of providing feedback to customers waiting in line—and credits work done at Disneyland (and later Disney World), where systems are in place that let people know how much longer they have to wait in line (Disney often overestimates the expected waiting time, so people are pleased when the actual time is less). At the same time, entertainment is often offered to the people waiting in line. It is also important, he notes, that the line is moving, so the people waiting perceive that they are making progress. The positive

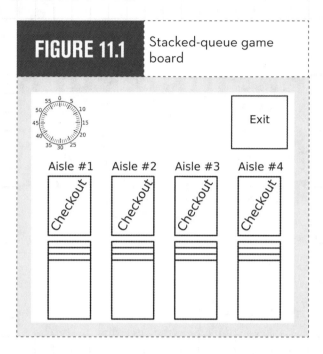

FIGURE 11.1 Stacked-queue game board

aspect of feedback is also why airline pilots usually announce how much longer they estimate they will be waiting in the taxi line before taking off, because it has been found that passengers are less upset about waiting when provided such feedback regarding their progress (Larson 1987).

Investigating Queuing Theory Materials (Teacher Background Information)

Materials

Have students work in groups of three for this simulation lesson. For the Engage phase, each group will need paper and pencils for making a sketch of some type of line in which people wait. In the Explore phase, each group of students will need two game boards (see Figure 11.1 for the stacked game-board example), one standard deck of 52 playing cards, and nine copies of a clock face (see Figure 11.1 for clock and *www.nsta.org/more-engineering* for a reproducible PDF of clocks). Finally, each group will need paper for making data tables (see Table 11.1) and a sticky note to make a label.

TABLE 11.1   Data table (in minutes) for a simple stacked checkout system

Aisle #1	Aisle #2	Aisle #3	Aisle #4	Elapsed time
3	7	2̶	12	0
(3 − 2) = 1̶	(7 − 2) = 5	5	(12 − 2) = 10	2
6	4	4	9	3
2			5	7

Engage

This lesson should commence with students sharing their experiences waiting in lines—perhaps at school, the movies, supermarkets, or department stores. Consider some of the places students wait in line at your school: for buses, the cafeteria, or the library. Does the cafeteria, for example, serve from one or two lines? After the discussion, students should make a sketch of one type of queue, demonstrating how it moves people through the line. This visualization should help prepare them for the simulation that follows.

Explore

Discuss the two common types of lines used in most retail establishments: a stacked system with a finite number of checkouts, each with its own line, or all lines combined into one serpentine line, in which the first person moves to the next available checkout. Prior to beginning the exploration, students should predict whether there will be any difference in the two systems and, if so, which system might be more efficient.

The queuing simulation presented here is quite simplified, but still takes a bit of practice. We suggest that teachers run through the simulation themselves prior to teaching the lesson. To begin, shuffle and then select 20 random cards. The cards represent customers, and the numbers on the cards represent the time (in minutes) it takes for that particular customer to be serviced at the checkout (count face cards as

FIGURE 11.2   Stacked queue, initial deal

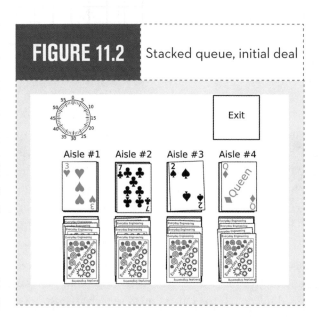

11, 12, and 13, respectively). For the first model, a simple stacked system, you will have four checkouts. Deal the 20 cards as shown in Figure 11.2. Data will be collected in a table similar to the one shown in Table 11.1. Record the numbers in each of the aisles on the data table as shown, in this case, 3, 7, 2, and 12. Note that the customer with the lowest number is the 2 in Aisle #3. That customer will require two minutes to be serviced. This customer will be moved to the exit, and two minutes will be recorded in the data table and on the clock face (see Figure 11.3, p. 86). This is represented by drawing a line through

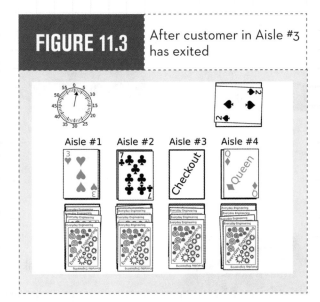

FIGURE 11.3 — After customer in Aisle #3 has exited

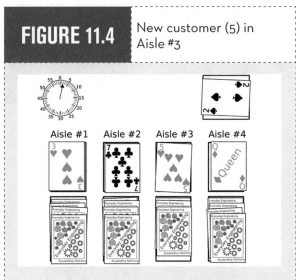

FIGURE 11.4 — New customer (5) in Aisle #3

the 2 in Aisle #3 of the data table (shown in red in Table 11.1, p. 85, for emphasis). Since two minutes have also passed for the other three customers at the checkout, subtract 2 from each and record the result in the row below (also shown in red in Table 11.1). They will now read 1, 5, empty cell in Aisle 3, and 10.

Now place the next customer in line in Aisle 3 at the checkout, in this case, a 5 (as shown in Figure 11.4 and in green in Table 11.1). From this point on, you need to consult the data table to see which customer will be next to finish the checkout process. Here we find that the 3 in Aisle #1 has just one minute remaining. In the next row, add one minute to the elapsed time, which is now three minutes. Move the customer in Aisle #1 to the exit and indicate this in the data table by crossing out the 1 in Aisle #1. Add a minute to the clock as well, indicating the three minutes of elapsed time. Subtract one minute from the customers at the checkout in Aisles #2, #3, and #4, which now should read empty cell in Aisle #1, 4, 4, and 9. Finally, place the next customer in line in Aisle #1 at the checkout, in this case, a 6 (shown in black type in Table 11.1).

Looking at the data table, you can see that in four minutes, both the 7 in Aisle #2 and the 5 in Aisle #3

will have completed the checkout process. Add four minutes in the time column and cross out both of those customers, change the clock and move both customers to the exit. Subtracting four minutes from the customers already at the checkouts leaves a 2 in Aisle #1, empty cells in Aisles #2 and #3, and a 5 in Aisle #4. Move the next two customers (in this case, a 4 and a 7) in Aisles #2 and #3 to the checkout (see Figure 11.5).

Looking at the game board, note that there are now a different number of people waiting at the checkouts (three each in Aisles #1 and #2, only two in Aisle #3, and still four in Aisle #4). This is because the first person in Aisle #4, a queen, requires 12 minutes to be checked out. (Hopefully, you are not in Aisle #4 behind the queen, because her line is certainly starting out the slowest!)

Continue the simulation and data recording until all customers have been moved through the checkouts. Note that some of the aisles will run out of customers before others. For this model, all customers should remain in their original aisles. Note the total time for all customers to complete checking out.

Figure 11.6 shows the streamed model setup with the appropriate game board. Shuffle the same 20 cards as used above. You will need a new clock to record

FIGURE 11.5	New customers (4 and 7) in Aisles #2 and #3

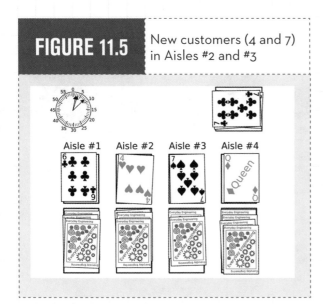

FIGURE 11.6	Streamed queue setup with first customers at checkouts

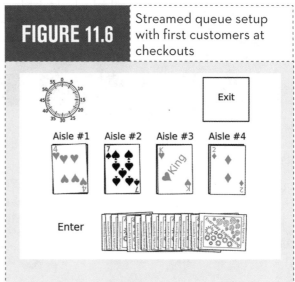

the elapsed time. Note that this setup still has four checkouts but just one line. The first four customers have been placed at the checkouts and the remaining 16 are all in one line. When one customer completes the checkout process, the next in line takes that position. Otherwise, the simulation is played in the same manner as the previous one.

To test the express-lane model, have students remove the five lowest cards from the pack of 20 they have been using. The five lowest cards represent the customers for Aisle #1, which is now deemed the express lane on the stacked game board. Students can label the express lane with a sticky note. Students should deal the remaining 15 cards equally among the other three aisles and conduct the simulation as described above. The customers will move through the express lane rapidly, but students must continue the simulation until all 20 customers have checked out.

Explain

Students will combine the data collected (Table 11.1) from each of their three trials to construct a table like the one in Table 11.2 (p. 88). As our sample data in Table 11.2 show, the streamed queue is generally faster

than the stacked one. Note that since the cards (customers) are distributed randomly, you may have an instance where this is not the case; that is, once in a while, the stacked may be faster than the stream. When you find class averages, however, the streamed system will be faster. Furthermore, you can see from our data that the time for the express lanes are significantly less than the remaining three stacked aisles.

In the discussion above, we have considered the total time for all customers to complete the checkout process. If you wish to have students do additional data analysis, you might also consider finding the average time each customer has to wait in line (to simplify the analysis in this simulation, all customers arrive in line at the same time). Dividing the total time by the number of customers results in the average time for the *system* to process *all* customers and *not* the average time each customer waits in line. To determine the average waiting time for each customer, one must add the time each customer waits to check out (in other words, the sum of the elapsed-time column) and then divide by the number of customers. Note that it is likely that more than one customer may have the same elapsed time, and in that case, each one must be added to

TABLE 11.2 Typical data compilation

Type of line	Total time (min.)			Average time of three trials (min.)
	Trial 1	Trial 2	Trial 3	
Stacked (*n* = 20 cards)	59	44	53	52.0
Streamed (*n* = 20 cards)	46	40	41	42.3
Express (*n* = 5 cards)	11	14	12	12.3
Stacked, non-express (*n* = 15 cards)	49	50	47	48.7

the total. For example, for the time period shown in Table 11.1 (p. 85), one customer checked out in two minutes, another in three, and two customers checked out after an elapsed time of seven minutes. Thus, the average time each customer stood in line is 2 + 3 + 7 + 7 = 19/4 = 4.75 minutes.

Extend

After students review and identify their examples from the Engage phase, direct the discussion toward other types of queues they may encounter. Essentially, any place where someone (or something) must wait its turn for processing involves a queue. Most students will recognize retail establishments and school lines as examples of queues. You may wish to provide photos of other examples to stimulate students' thinking about queues. Consider toll-road collection booths. Most are of a stacked design but have special lanes for cars with electronic transponders to pay the toll without using cash or having to stop, similar to an express lane in a store. In addition, there must be booths for cars that stop and use cash. Engineers designing such a system must be aware of how many cars are likely to pass through with or without an electronic transponder, as well as how the amount of traffic will vary as a function of the time of day

(or day of the week), and then adjust how many toll booths need to be open. Designers must also consider the costs incurred by the electronic booths and the wages for operators in the cash booths. Grocery store managers must consider many of the same factors—how customer load varies with time of day (or day of the week), the times that people buy few items (requiring more express lanes), and the times that customers are likely to purchase more items. In both the grocery store and the toll booth situations, those in charge must consider optimum customer service at some reasonable cost.

Other types of queues include transportation examples, such as the movement of traffic on a highway (as a function of the number of lanes and amount of traffic), air-passenger security screening, and airplanes queued for takeoff or landing. When you telephone for assistance, you will likely be routed through a call center, where you may have to wait for the next available operator. Some provide a menu of different services so you can select a separate queue to wait in for more specialized assistance.

Most of the queues we have been discussing provide service to the next person waiting in line—that is, people are waited on essentially in the order in which they arrive. This is not always the case. Consider an emergency room at the hospital: There, one usually finds a triage nurse whose job is to determine which patients are in more critical condition and therefore need to be treated before others. Another method where the next in line is not necessarily the next to be served is the loading of airplanes; usually the airline's "best" customers are allowed access to the plane first.

Evaluate

Student simulations should readily show that the streamed system is more efficient.

Conclusion

Engineering shows up in unexpected places, including waiting in line at the grocery store or for a table at a restaurant. You may now realize why you are usually in (one of) the slowest lines at the store: If there are four stacked lines, you have only a one in four (25%) chance of being in the fastest line.

References

Larson, R. C. 1987. OR Forum—Perspectives on queues: Social justice and the psychology of queuing. *Operational Research* 35 (6): 895–905. *http:// pubsonline.informs.org/doi/pdf/10.1287/opre.35.6.895*

NGSS Lead States. 2013. *Next Generation Science Standards: For states, by states.* Washington, DC: National Academies Press. *www.nextgenscience.org/ next-generation-science-standards*

Plus Math. 1997. Agner Krarup Erlang (1878–1929). *+Plus Magazine. https://plus.maths.org/content/ os/issue2/erlang/index*

Resources

"Dr. Queue" helps you avoid rage in line. *www.npr.org/ templates/story/story.php?storyId=120769732*

Find the best checkout line. *Wall Street Journal. www. wsj.com/articles/SB10001424052970204770404577 082933921432686#project%3DLINES120811%26a rticleTabs%3Dinteractive*

Stordahl, K. 2007. The history behind the probability theory and the queuing theory. *Telektronikk http:// titania.ctie.monash.edu.au/netperf/Stordahl HistoryOfProbQueueTheory2006.pdf*

Queuing theory video. EngineerGuy.com. *www.engineerguy. com/videos/video-lines.htm*

Queuing video. *www.youtube.com/watch?v=_CBD2z51u5c*

ACTIVITY WORKSHEET 11.1 Investigating Queuing Theory

Engage

1. How often do you have to stand in line to wait for something? Brainstorm with your group the different places you find yourself waiting in line. Record your ideas and share with the class.

2. What are some different ways that lines operate to move people through more quickly? Review the class list of different kinds of lines again. Discuss and then make a group sketch modeling how at least one of the lines functions to move people efficiently. Share and discuss these models with the class.

3. In this exploration, you will use a simulation to test different systems of waiting in line to determine the efficiency of each.

Explore

1. Although there are many variations, there are two basic checkout lines that stores use. Most grocery stores have a line for each clerk. Some other stores use just one line that directs people to the next available clerk. Predict which model is the most efficient. Explain your reasoning.

2. In this simulation, you will use 20 standard playing cards, a game board that represents the different models of checkout lines, and a simulated clock to record the amount of elapsed time it takes all 20 people (cards) to check out and move out of the system. Each group will require three copies of a data-recording table. Your teacher will provide the materials required for the simulation.

3. In the first model, simulating most grocery stores, you will have four checkout lanes with five people

(cards) in each line. Your teacher will explain how to conduct the simulation and collect your data.

4. In the second model, which is similar to the one used in some department stores, you will have four checkout lanes, but just one line of 20 people. As a checkout becomes available, the next person in the line moves forward. Your teacher will again explain how to conduct the simulation and collect your data.

5. In the third model, you will test a variation of the grocery store model, in which one of the four checkout lanes will become an express lane. Your teacher will again explain the model and how to collect data.

6. Return the cards to the deck, shuffle, and deal 20 new cards.

7. Repeat steps 3–5 to collect data for a second trial of your test. Record your data.

8. Repeat steps 6 and 7. Record your data for the third trial of your test.

Explain

1. Analyze the data you have collected and record each trial in a table like the one shown on the next page. What was the total amount of time it takes for all of the people to check out of each different model? What was the average time of the three trials for each model?

2. How do the average times of stacked versus streamed compare? Which is more efficient?

3. How do the express versus the stacked, non-express times compare? Why might someone with few purchases prefer this model of checkout?

Type of line	Total time (min.)			Average time of three trials (min.)
	Trial 1	Trial 2	Trial 3	
Stacked (n = 20 cards)				
Streamed (n = 20 cards)				
Express (n = 5 cards)				
Stacked, non-express (n = 15 cards)				

Extend

1. Review the examples of lines (called *queues*) you discussed in the Engage stage and identify the type of system, stacked or streamed, for each case.

2. Where have you encountered express lanes?

3. Think of additional examples of queues and identify the type.

4. In this activity, we have considered only the time for 20 customers to check out. If you were a store owner, what other factors might you consider to determine the type and number of checkouts?

Evaluate

1. There are times when some customers require additional time at the checkout—perhaps an item will not scan and its price needs to be verified. Consider how each system, stacked or streamed, might be best suited to accommodate such situations. Explain your reasoning.

2. Conduct another simulation in which you randomly select 18 cards along with the two Jokers. The Jokers will have a value of 25 minutes. Record at least one round of data for each model you test. Using your data as evidence, support your conclusion regarding the best system for situations in which customers may require additional checkout time.

PART 4

Engineering Ordinary Things

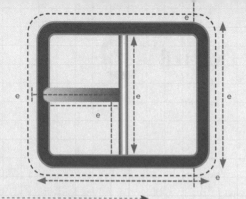

CHAPTER 12

KEEPING IT TOGETHER— FASCINATING FASTENERS

ONE, TWO, BUCKLE my shoe. Three, four, there are so many more—fasteners, that is. Ask people to name fasteners, and you will get a number of different answers. To a carpenter, fasteners might be nails and screws; to a metal worker, rivets and bolts; to a seamstress, buttons and snaps; and to a child, building blocks and Velcro. Think of the myriad fasteners people use to attach clothing and jewelry items—everything from buttons to clips, clasps, and hooks. Furthermore, there are numerous designs of each of these types of fasteners. For our purposes, we will define a *fastener* as a piece of hardware that connects two items; is relatively easy to connect and disconnect; and, when connected, tends to stay connected. Fasteners for jewelry and clothing are intended for repeated uses rather than one use, such as adhesive tape, or relatively permanent attachment such as glues.

In this 5E learning-cycle lesson, students will construct and test a fastener of their own design and compare their design to those of their classmates. Students will evaluate their original design in light of alternative criteria. Then, students will redesign their original plan and perhaps test the fastener again. The *Next Generation Science Standards* (NGSS) suggest that middle-level students "[e]valuate competing design solutions using a systematic process to determine how well they meet the criteria and constraints of the problem" (MS-ETS1-2; NGSS Lead States 2013). The topic of fasteners relates to the *NGSS* disciplinary core

idea of Structure and Properties of Matter: "Structures can be designed to serve particular functions by taking into account properties of different materials, and how materials can be shaped and used" (MS-PS1-3).

Historical Information

Early humans held their clothing together with pins they fashioned from sharp animal bones. The use of buttons is basically as old as clothing. Prehistoric people used buttons made of materials such as shells, wood, and bones to decorate and fasten their clothing. As time went on, buttons became more of a fashion statement, often made of metal (gold and silver for wealthy individuals) (Sewing Mantra 2010).

Metallic-snap clothing fasteners were introduced in the late 1800s in Europe. Initially, snaps found a market in actors' costumes, allowing for quick changes between scenes. By the 1930s, snaps were introduced as a fastening option for home sewers. At about the same time, they revolutionized infant clothing with the introduction of the snap-closure onesie, providing a nightgown alternative.

Another common type of fastener is the simple hook and loop. Many jewelry items are constructed with this type of closure. The most famous of all hook and loop fasteners is Velcro, which was invented in 1941 by Swiss engineer Georges de Mestral after he noticed burrs that attached to his pants and dog after a walk in the woods. He named his invention after the French words for velvet and crochet. Velcro is a popular example of biomimicry. The U.S. Army found that Velcro on soldiers' uniforms was not functioning properly when exposed to the sand of desert warfare. Ultimately, they solved the problem of nonfunctioning hook and loop fasteners with something much older—buttons (Suddath 2010).

Investigating Fasteners (Teacher Background Information)

Materials

For this investigation, it is best to have jewelry wire, which can be purchased at a hobby or craft store. It is likely to be called memory wire because it tends to return to its original shape. Thus, it can be used to make the simple springs needed to construct clasps and fasteners. Ordinary wire that you may have lacks this springiness and is therefore not suitable for many of the fasteners shown here. We are using 24-gauge wire that is often sold by the ounce, where 1 ounce will yield approximately 10 m of wire for about $7. If you have five classes with six groups in each, you will likely need two rolls of wire.

Have students work in groups of four for this investigation. In the Explore phase of the lesson, each group will need two pieces of memory wire, the first about 15 cm long and the second between 40 and 50 cm. Each group also will require a pair of needle-nosed pliers. Round-nosed pliers (for jewelry making) are preferable, but not necessary. Pliers are available online for $3–$4 a pair. Additional wire will be needed if you choose to have students actually construct and test their designs for the Evaluate phase. In addition, you will need to have a collection of assorted fasteners such as snaps, hooks and eyes, buttons, safety pins, Velcro, and so on, available for students to observe and classify in the Extend phase. A hand lens for each student will also be helpful for observation. You may also wish to provide images downloaded and printed from online sources.

Engage

While fasteners are ubiquitous to daily life, most of us rarely think about how they function, and even fewer consider their engineering and design. Given the incredibly wide array of different types of fasteners, in this lesson, we focus primarily on those used in clothing and jewelry. Students' prior knowledge regarding fasteners will vary, but may include ideas such as "easily connected and disconnected, but tend to stay connected unless intentionally unfastened." This is in contrast to other fasteners, such as those used in construction, which are often designed not to be disconnected—nails, rivets, and dowels, for example.

Explore

Safety note: Students must wear safety goggles for the duration of the exploration and any time they are working with the wire. Students need to be cautious about being pricked with the ends of the wire, bending the wire, and using the pliers.

There are two parts to the Explore phase. The first is to familiarize students with the pliers, wire, and how a simple hook-type fastener functions. To that end, have students construct a simple fastener at the outset (see Figures 12.1 and 12.2). The second part is the engineering challenge, which is to design and build a fastener to hold two items together. The memory wire can be formed into a variety of shapes to make numerous designs of fasteners (see Figure 12.3).

Engineering designs usually must satisfy many criteria. They may address the safety, use, durability, production, or cost of the design. For example, you might want to make a fastener for a parachute that, regardless of cost, can be relied on to have an extremely low probability of failure. On the other hand, a fastener for an inexpensive trinket might be designed so that it can be manufactured inexpensively with less reliability in terms of possible failure. In our challenge, the engineering criteria are that the fastener must be easily connected and disconnected, and that it stay

FIGURE 12.1 Forming wire with pliers

FIGURE 12.2 Simple wire fastener, closed and open

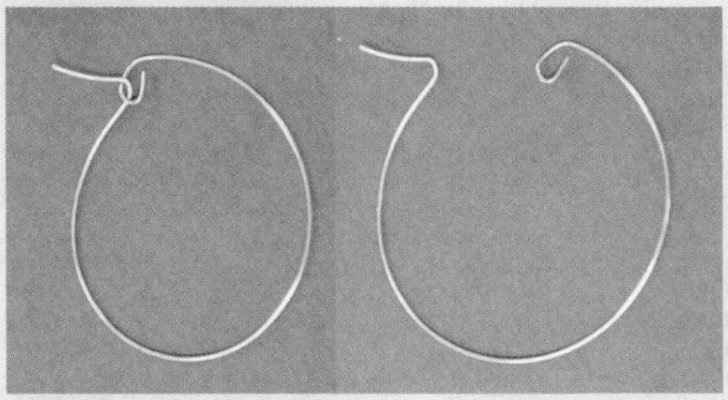

FIGURE 12.3 Different fastener designs

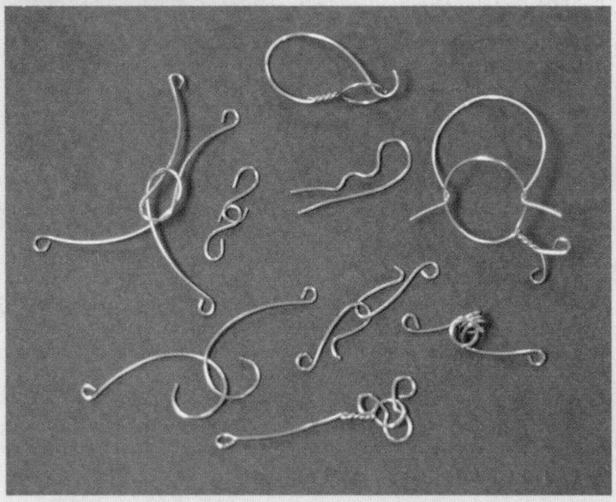

connected for at least 10 seconds while holding the weight of the object being connected.

All students should have the opportunity to design and sketch their own fastener to meet the engineering challenge. However, if you are not able to supply more than one pair of pliers per group, then each group will need to select one of their designs to construct. Students will not be able to bend this wire with their fingers. If you have access to more pliers, it would be ideal to have students make their own fasteners.

Explain

Student groups will need to analyze their fasteners to determine if they met the design criteria. They should also be able to describe how each of their group's fasteners works. Through discussion, have each group report the analysis of each design to the class. Have students classify the designs into different groups as a function of how they operate. For example, one group may produce loops and hooks. Another group's fastener may work by squeezing together a larger part so that it slides through a smaller opening. When released, it remains in place. Other types might include S-hooks or possibly even a snap. Generally, these fasteners stay together because of either the tension caused by the spring of the wire or, in the case of hooks, the gravity pulling the hook down on the loop. Others, such as the screw types, become entwined or twisted together. The memory wire was selected for the material to be used because of its property of holding its shape and springing back to its shape. Other wire does not exhibit this property. If time allows, an extension could be for students to try other types of wire to experience the properties of different materials.

FIGURE 12.4 Assorted clothing fasteners

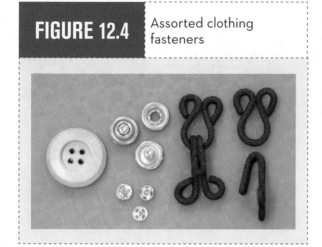

FIGURE 12.5 Close-up view of a snap socket

Extend

Now that students have designed their own fasteners, the Extend phase provides them with the opportunity to look at a variety of clothing fasteners such as buttons, snaps, and hooks (see Figure 12.4). Typical buttons are usually round and fit through a slit-shaped opening that needs to be slightly longer than the diameter of the button. To unfasten a button, it must be turned so that it can be pushed back through the slit. If the slit is too long, the button will not stay closed because the total area of the opening is greater than the area of the button.

Another type of fastener is a snap. Most snaps use a post and a socket that snap together because the post is slightly larger than the socket. Usually, the socket is designed with a small amount of flexibility to accommodate the larger post. Thus, they remain fastened when forced together until sufficient force is again applied to separate the post from the socket. Some designs have small slits while others have springs to provide this flexibility (see Figure 12.5 to view the two straight bars that are part of the springs). Other snaps work because of a flexible post rather than a flexible socket. They may be made of several pieces that

collapse slightly and then spring back, which allows them to remain fastened in the socket. Plastic snaps may function this way.

To continue with this extension activity, students can create a fastener inventory of items within the classroom or as a homework assignment. Another extension to the lesson could focus on the historical aspect of clothing fasteners. Students could research clothing and fasteners from different time periods or different cultures. Finally, students could research other examples of biomimicry.

Evaluate

Because engineering is an iterative process, it is good to change the design criteria after students have created their initial fastener. There are multiple possible design solutions to address the criterion of increased weight. Some of these include using thicker wire or increasing the number of strands of wire. Other students may choose to reconsider their fastener's closing system with a modification or a completely new design. This provides students the opportunity to apply the new knowledge they have acquired through the lesson.

Conclusion

Engineering is indeed fascinating. Many of us do not usually give much thought to how the buttons or snaps on our shirts stay fastened; but, once again, we see that someone had to design them. Without this innovation, might we still be pinning our clothes together with sharp pieces of bone?

References

NGSS Lead States. 2013. *Next Generation Science Standards: For states, by states.* Washington, DC: National Academies Press. *www.nextgenscience.org/next-generation-science-standards*

Sewing Mantra. 2010. A brief history of buttons—Sewing button origins. *www.sewingmantra.com/index.php/sewing/sewing-button-history*

Suddath, C. *Time.* 2010. A brief history of: Velcro. June 15. *http://content.time.com/time/nation/article/0,8599,1996883,00.html*

CHAPTER ⑫

ACTIVITY WORKSHEET 12.1   Investigating Fasteners

Engage

1. Brainstorm with your group a list of ways people use fasteners. Share this information with your classmates.

2. What kinds of fasteners are used on clothing and jewelry? Make a list to record your ideas. Share with your classmates. What characteristics do most of these fasteners have in common?

3. In this investigation, you will design, build, and test a fastener to hold two items together.

Explore

Safety note: You must wear safety goggles for the duration of this exploration. Follow teacher directions for use of pliers and metal.

1. Think about how you might design a simple method that easily hooks and unhooks the ends of a piece of wire together. Make a sketch of your idea. Discuss the designs with your group and decide which one you wish to make, using the materials provided by your teacher. Each group should demonstrate to the rest of the class how its fastener hooks and unhooks.

2. The engineering challenge of this exploration is to design and build a fastener to hold two items together. For example, you might design a fastener that will attach some keys to a backpack. The criteria to meet this challenge are that the fastener must be easily connected and disconnected, it must not disconnect on its own, and it must be able to support the weight of the item.

3. First, each member of your group should make a sketch of his or her design.

4. As a group, discuss each of the designs and how well each of them may meet the design criteria.

5. Select at least one design to construct and test. State why you chose this design or what you think will make this design succeed.

Explain

1. How well did the design meet each of the criteria? Give reasons to explain why they did or did not meet the criteria. Record your observations.

2. Describe how the fastener connects and can be disconnected. Explain why it stays connected.

3. Share the designs and analyses with your classmates.

4. Classify the designs into different types based on how they work. Are any of them similar to clothing and jewelry fasteners the class previously listed? Explain.

Extend

1. Look at the fasteners your teacher has provided. Perhaps you can find some additional examples online or at home.

2. Analyze how they connect, stay connected, and disconnect, and record your observations. Which materials were specifically used in each fastener?

3. Classify these fasteners based on how they work. Record your ideas.

Evaluate

Imagine one criterion has changed and the fastener you built must support a wide variety of items of different weights. Redesign your group's fastener to meet this new criterion. Keep in mind that you still must meet the other criteria—the fastener must hook and unhook easily and remain closed when fastened. Make a sketch and write a paragraph describing the changes you plan for your design. What property of the fastener material will be the most desirable?

CHAPTER 13

TWISTING AND BRAIDING—FROM THREAD TO ROPE

PEOPLE HAVE BEEN twisting plant fibers together to make cords and ropes since prehistoric times. Over time, the technology has exploded, giving us literally hundreds of types of threads, strings, yarns, ropes, and cables made from dozens of different natural as well as synthetic materials. The size range of stringlike products or cordage is very broad: from the microfibers used in towels, which are about one-hundredth the diameter of a human hair, to the steel cables in the Golden Gate Bridge, which are nearly a meter thick. A new material currently being investigated, *graphene*, is made of a single layer of carbon atoms in a honeycomb pattern. A string made of graphene is the diameter of a very sharp pencil and could lift a grand piano (Dunn 2013).

In this 5E learning-cycle lesson, students will design a procedure to test the breaking point of threads made from various materials. These activities will help students understand the crosscutting concept Structure and Function from the *Next Generation Science Standards*: "Structures can be designed to serve particular functions by taking into account properties of different materials and how materials can be shaped and used" (MS-PS1-3; NGSS Lead States 2013). To this end, we could use a thin steel cable for sewing clothing, and though it would be quite strong, it would lack other properties desired of sewing thread, such as light weight, softness, and the ability to be dyed different colors. Students will also look at a variety of cordage to observe different structural designs of threads, ropes, and cables.

Historical Information

Humans have been making ropes and cords for thousands of years. Fossilized pieces of rope have been discovered that are more than 15,000 years old. Many believe that people have been using ropes and tying knots for a much longer time than this, perhaps even predating the use of fire in 400,000 BCE: "It is likely that the earliest 'ropes' in prehistoric times were naturally occurring lengths of plant fiber, such as vines, followed by the first attempts at twisting and braiding these strands together to form the first proper rope in the modern sense of the word" (De Decker 2010, p. 2).

In rope making, fibers are spun into yarns, which are twisted into strands that are then twisted again (in the opposite direction) into rope. Twisting the strands in the opposite direction is designed to prevent the rope from unwinding and coming apart. This basic process has been in place since about 1500 BCE in Egypt and then was refined during the Middle Ages as more uses for rope developed, especially in sailing vessels. The innovation of wire rope occurred in the second half of the 19th century for use in mines. In the mid-20th century, the advent of synthetic fibers such as nylon and polyester resulted in the decline of ropes made of natural fibers.

The history of thread follows a similar path. Because thread is not easily preserved, early evidence of its use is the discovery of sewing needles made from bones dating back to around 17,500 BCE. Early threads were fashioned from linen and wool. Later, around

350 BCE, the Chinese began spinning silk into thread and sometime later exported it to the Europeans. The processing of cotton came considerably later—in Arab countries in the fourth to fifth centuries CE. As with ropes, the development of synthetic materials revolutionized the thread industry, beginning in the 1940s and 1950s (European Federation of Sewing Thread Industries, n.d.). You may wish to have students investigate some of the history of cordage on their own or perhaps coordinate with the social studies department.

Investigating Stringlike Materials (Teacher Background Information)

Materials

Throughout this activity, you will need to provide several types of threads, ropes, and cables. We recommend you search your sewing kit, junk drawer, and garage before purchasing anything. In the Engage phase, each student will need a short piece of cotton thread, a piece of string, a piece of dryer lint (about the size of a nickel), and a hand lens (5× magnification is optimal).

In the Explore phase, each group will require at least three types of thread—perhaps 100% cotton, 100% polyester, and 100% nylon. These three will exhibit a broad range of strength (and stretching) characteristics. In addition, we recommend each group also observe or test three additional samples. There are many others available, such as blended cotton/polyester, nylon/polyester, metallic, quilting, machine embroidering, and silk threads. To test the variable of thickness, you will need two threads of similar composition but different thicknesses. Each group will require three 60 cm lengths of each thread type. One spool of thread should be more than sufficient for six classes with six groups of students. Thread of this type is available at fabric stores for $1 to $2 per spool, with specialty threads costing somewhat more.

Materials will vary if you have students design their own method of testing. To conduct the thread-strength testing as described, each group of four students will

FIGURE 13.1 Setup with plastic bucket hanging from broom handle

need one small plastic pail (a standard 2.5-gallon pail will work well), ten 500 ml water bottles, a broom handle (or other support from which to hang the bucket; see Figure 13.1), duct tape to secure the broom handle, scissors, and a meter stick. To reduce the number of bottles needed, you may opt for 1 and 2 L bottles. Students will need a balance to find the mass of their empty bucket and bottles, or you may wish to simply provide this information to the entire class.

In the Extend phase, each group of four students will need three or four samples (perhaps about 30 cm) of rope and cable. One should be an example of a braided rope over a core, such as a clothesline, and the other should be a twisted type, such as polypropylene or nylon ropes. Ropes should be taped a few centimeters from the end to keep them from fraying excessively. Ropes are available at most hardware stores for 29–59 cents per foot. Because the rope is reusable, for six classes of six groups, a total of about 2 m of each rope will be needed. Thicker nylon ropes are generally more expensive. Note that we have provided strength and composition information for sample ropes and cable (Table 13.1). This information is usually on the spool label when you purchase the materials from a hardware store or

TABLE 13.1	Rope and cable strength comparison		
Thickness	**Composition**	**Arrangement of strands**	**Working load (newtons)**
1/4"	Polyester	Twisted	663 (149 lb.)
3/8"	Polyester	Twisted	1,486 (334 lb.)
1/4"	Polypropylene	Twisted	418 (94 lb.)
3/8"	Polypropylene	Twisted	903 (203 lb.)
1/4"	Cotton	Twisted	214 (48 lb.)
3/8"	Cotton	Twisted	481 (108 lb.)
1/4"	Steel	Twisted	4,895 (1,100 lb.)
3/8"	Steel	Twisted	10,858 (2,440 lb.)
1/4"	Polyester	Braided	712 (160 lb.)
3/8"	Polyester	Braided	1,736 (390 lb.)
1/4"	Polypropylene	Braided	668 (150 lb.)
3/8"	Polypropylene	Braided	1,113 (250 lb.)
1/4"	Cotton	Braided	334 (75 lb.)

Note: Table adapted from Bevis Rope 2010 and Engineering ToolBox (a) and (b).

can be found on most rope manufacturers' websites (see References [p. 108]).

Engage

Begin by initiating a discussion of what students think dryer lint is and where it comes from. Then ask students if they have heard of people spinning wool into yarn and how that process might work. Dryer lint is a collection of short fibers that are removed from fabrics during the drying process; these fibers are easily pulled apart because they are just matted together on the lint trap. However, if you twist some of the lint in one direction, you will form a stringlike material with significantly more resistance to pulling apart (see Figure 13.2).

FIGURE 13.2	Dryer lint, twisted and untwisted

| **FIGURE 13.3** | Various threads × 20: (A) Cotton, (B) Spun polyester, (C) Nylon, (D) Metallic |

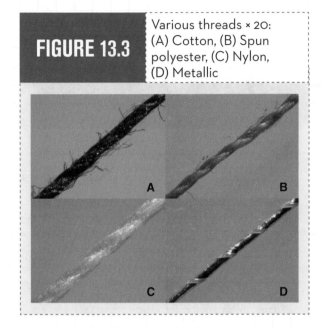

This is essentially what happens when wool is spun into yarn: The short fibers are twisted together to form one continuous strand. Many students will have the misconception that a piece of thread is similar to nylon fishing line—that is, one long continuous piece, or monofilament, of material. The experience of twisting the lint will help students understand that strings, threads, and ropes made of natural fibers are not monofilaments but made of many shorter fibers that

are twisted together. Also, many synthetic (or metallic) stringlike materials are made by twisting thin but long continuous strands together (see Figure 13.3). Polyester threads are sometimes made from short fibers similar to cotton or from long, twisted filaments.

Explore

Students will see some similarities among the three threads—all are made by twisting numerous strands together. While students may not realize it, all three types can be dyed virtually any color. The amount of fuzziness and how they reflect light are two notable differences. Some students may also notice differences in the texture, as the cotton and polyester have a smooth feel to them. Finally, your samples may vary as to their thickness (see Table 13.2)

While you can have students design their own testing procedure, one method follows. In this exploration, students either determine the breaking point of different compositions of threads or whether the thickness of a thread affects strength. Of course, you may wish to have each group of students test both variables. One method is shown in Figure 13.1 (p. 102), where a thread is attached to a support on one end and a plastic pail on the other. Note that spring scales are difficult to read at the breaking point, and the scales typically found in school labs are too small for this

| **TABLE 13.2** | Thread observation table |

Thread sample	Composition of thread	Observations with hand lens	Other observations
#1	100% cotton	Twisted and fuzzy, made of smaller threads	Dull color, smooth
#2	100% polyester	Twisted, a little fuzzy	Somewhat shiny, smooth
#3	100% nylon	Not fuzzy, twisted	Slick, shiny

TABLE 13.3 Breaking point of different thread compositions

Composition of thread	Trial #1 (liters)	Trial #2 (liters)	Trial #3 (liters)	Average breaking strength (liters)
100% cotton	0.5	0.6	0.5	0.5
100% polyester	0.7	0.7	0.7	0.7
100% nylon	4.4	3.9	4.2	4.2

activity. We suggest that both ends be tied in double knots so that the ultimate failure is in the thread rather than the knot. Water bottles are added to the pail until the thread breaks. By varying the amount of water in the bottles, students can be reasonably precise in finding the breaking point. If one bottle has 100 ml, two have 200 ml, and the rest have 500 ml, students can determine the breaking point to within 100 ml, or 100 g.

Note that our pail is about 20 cm from the floor not only to minimize the thud when the thread breaks but also to allow for threads to stretch—the nylon one is especially prone to stretching. We also recommend that the broom handle or support bar be secured to the table or other support. Students should wear eye protection and be careful when the thread breaks and the bucket

falls. Students should create a data table (see Tables 13.3 and 13.4) that includes multiple trials. Check student procedures before they begin the exploration.

The observations of additional threads will vary, depending on what types you provide students. For example, if you provide a "metallic" thread, students will likely note that the metallic color is wrapped around a core material. Machine embroidery thread may be smoother than other polyester threads.

Explain

Students should find that the 100% nylon thread is much stronger than both the 100% cotton and the 100% polyester. In addition, the strength of the thread increases with increased thickness. In general, this is the

TABLE 13.4 Breaking point of different thread thicknesses

Thickness of thread	Trial #1 (liters)	Trial #2 (liters)	Trial #3 (liters)	Average breaking strength (liters)
Thin 100% polyester	0.7	0.7	0.7	0.7
Thick 100% polyester	2.9	2.2	2.5	2.5

case for any given material; that is, strength is a function of cross-sectional surface area. Nylon, a synthetic material, is stronger than the polyester and the cotton because of its molecular structure, which includes very strong bonds between its component parts.

To determine the force required to break each thread, students need to add the mass of the empty pail and bottles to the mass of the water and then convert this number to newtons. In our case, the pail had a mass of 160 g, and each empty water bottle was 9 g. For example, in one trial, 4.2 L of water was required to break the nylon thread. This mass included nine bottles of water (eight 0.5 L bottles, plus one containing 0.2 L). The total mass then includes the water plus the bottles, plus the pail:

$$\text{mass} = 4.2 \text{ kg} + 0.081 \text{ kg} + 0.16 \text{ kg} = 4.441 \text{ kg,}$$

which rounds to 4.4 kg. We can convert mass to weight using the following relationship:

$$\text{weight} = \text{mass} \times \text{acceleration due to gravity (g)}$$

So, the weight (in newtons) required to break the nylon thread is as follows:

$$\text{weight} = 4.4 \text{ kg} \times 9.8 \text{ m/s}^2 = 43.1 \text{ N}$$

If your students have difficulty calculating the force exerted on the thread, you could simply note the total volume of water in liters that resulted in the string breaking, although doing the calculations is a good review of the relationship between force (weight) and mass. This might be a good time to help students realize that whether they measured the breaking strength in liters of water or in newtons, the order in which the threads break remains the same.

Discuss circumstances where it might be desirable to use different types of thread when stronger seams

FIGURE 13.4

Samples of ropes: (A) Braided polyester clothesline, (B) Braided nylon over core, (C) Twisted polypropylene, (D) Twisted steel cable

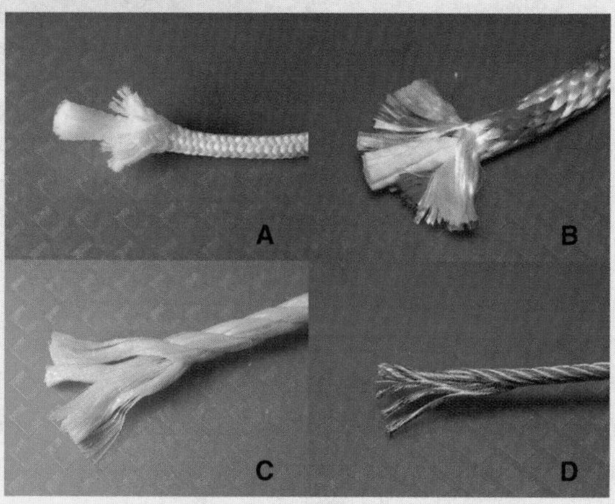

are required (e.g., in upholstery or parachutes). Nylon threads are clearly the strongest but are also prone to stretching. In addition, in some applications, one might desire a decorative or shinier thread. While we are not considering it here, some threads are more colorfast or resistant to sunlight than others (you may wish to investigate some of these variables as well).

Extend

Safety note: Students must wear safety goggles. Provide each group of students with three or four samples of ropes and cable as shown in Figure 13.4. Students should notice two general arrangements of the strands, twisted or braided. Twisted ropes may have a varied number of strands, but many common ropes are made from three strands. Note, however, that each of the strands is likely made up of many individual, thin strands called *yarns*. Braided ropes are either hollow or formed around an interior core. Each strand of the braid is also composed of many smaller yarns.

You will need to either obtain the strength of the rope and cable samples you have acquired or use the data we provide in Table 13.1 (p. 103). (*Note:* American teachers will find only ropes sold in thickness measured in inches and working loads in pounds.) Students should conclude that the steel cables are the strongest and the cotton ropes the weakest. For any material and construction, thicker cordage is stronger, and braided is somewhat stronger than twisted.

The strength of the ropes in Table 13.1 is recorded as the working load, which is 20% (or one-fifth) of the breaking point for each sample. This is to provide a safety factor so that ropes and cables are not used at their limits of structural integrity. Thus, to find the likely breaking point of any of the cordage, simply multiply the working load by 5. Therefore, we would predict that the ¼ in. steel cable will have a breaking point of 24,475 N (4,895 N × 5).

To determine the number of bottles of water required to break the clothesline in step 4 of the Extend phase on Activity Worksheet 13.1 (pp. 109–110), students must first change the force in newtons to kilograms of water. Because the weight (or force) is found by the equation $W = mass \times g$, we can solve for mass:

$$m = W/g = 334 \text{ N}/9.8 \text{ m/s}^2 = 34.1 \text{ kg}$$

Because the working load is 20% of the breaking point, we must multiply this mass by 5:

$$\text{breaking point} = 34.1 \text{ kg} \times 5 = 170.5 \text{ kg}$$

To determine the number of bottles of water, we must first subtract the mass of the empty bucket:

$$170.5 \text{ kg} - 0.16 \text{ kg} = 170.34 \text{ kg}$$

Finally, since each bottle of water has a mass of 0.509 kg (the mass of the water plus the empty bottle), we divide 170.34 by 0.509:

$$\text{number of bottles for clothesline}$$
$$\text{to break} = 170.34/0.509 = 334.7$$

(In actuality, we would need to use a much larger bucket!)

Evaluate

We have considered the strength of cordage only as a function of composition, construction design, and thickness. Students may have also noted that the nylon threads stretched more than the others during the exploration. There are, of course, many other characteristics to consider when deciding on the appropriate rope for a given function—cost, weathering resistance, and flexibility, to name just a few.

In this phase, we point out to students that polypropylene rope, while impervious to water, deteriorates when exposed to the ultraviolet radiation of the Sun. We ask students to study the information in Table 13.1 to determine what other ¼ in. ropes they could substitute for the polypropylene that would be of equal strength. They should be able to determine that both the twisted and braided polyester ropes and the steel cable would be of sufficient strength. They might infer (and you may choose to discuss with them) that the steel may be much stronger than needed for this purpose and likely more expensive; in time, it will likely rust as well. Due to cost factors, most clotheslines today are made of braided polyester or a cotton/polyester blend.

Conclusion

Cordage design is similar regardless of scale—threads, strings, ropes, and cables are often made by twisting thinner pieces together. As is frequently the case in engineering, once an idea has been invented, it often lasts, with innovations occurring over time. For cordage, strength is increased by the addition of more strands, and a convenient way to add strands is to twist them together. An innovation was to braid them, sometimes around a core material. It is interesting to note that braided fibers are usually made of thinner fibers twisted together. Readers of this column will note that reiterative innovation of an invention is a recurring theme—check valves in pumps, roller balls in pens, and spring design in clips and clamps (Moyer and Everett 2009a; 2009b; 2011).

Encourage your students to be aware of design innovation as they encounter their everyday world. Consider the design of your shoelaces. Most are still made of natural fibers rather than synthetic ones. Why? It is more difficult to keep knots in synthetic laces (and ropes) because they are slippery. Where else might examples of braided or twisted cordage be found?

References

Bevis Rope. 2010. Rope characteristics. *www.bevisrope. com/rope-info/rope-characteristics*

De Decker, K. 2010. Lost knowledge: Ropes and knots. *Low-Tech Magazine. www.lowtechmagazine. com/2010/06/lost-knowledge-ropes-and-knots.html*

Dunn, R. 2013. Element hunters. *National Geographic 223* (5): 112–23.

Engineering ToolBox (a). Polypropylene fiber rope—Strength. *www.engineeringtoolbox.com/polypropylene-rope-strength-d_1516.html*

Engineering ToolBox (b). Wire rope—strength. *www. engineeringtoolbox.com/wire-rope-strength-d_1518.html*

European Federation of Sewing Thread Industries. n.d. One of the oldest products in the world. *http:// eft-sewingthread.com*

Moyer, R., and S. Everett. 2009a. Everyday Engineering: What makes a Bic click? *Science Scope 32* (8): 38–42.

Moyer, R., and S. Everett. 2009b. Everyday Engineering: What makes a squirt gun squirt? *Science Scope 33* (2): 10–14.

Moyer, R. H., and S. A. Everett. 2011. Everyday Engineering: Clips and clamps. *Science Scope 35* (4): 16–21.

NGSS Lead States. 2013. *Next Generation Science Standards: For states, by states.* Washington, DC: National Academies Press. *www.nextgenscience.org/ next-generation-science-standards*

ACTIVITY WORKSHEET 13.1 — Twisting and Braiding—From Thread to Rope

Engage

1. Use a hand lens to look at the piece of lint your teacher has provided. Gently pull it apart. Now twist a sample between your fingers to make a stringlike fiber. Observe if it is now more difficult to pull apart, and make a sketch of your observations.

2. Compare a piece of cotton thread and a length of string to the lint and record your observations. How are they the same? Different?

3. In this exploration, you will design a procedure to test the strength, or breaking point, of different types of threads.

Explore

1. Compare the different samples of thread provided by your teacher. Construct a table like the one below to record your observations.

2. Based on the materials provided by your teacher, design a procedure to test the strength of each type of thread. Quantify the strength of the threads using the amount of water to reach the breaking point.

3. Create a data table to record your results for multiple trials for each type of thread.

4. After your teacher has approved your plan, conduct your testing. Wear safety glasses for testing.

5. Calculate the average breaking strength of each type of thread.

6. Observe the additional thread samples your teacher has provided. How do they compare to the ones you tested earlier?

Explain

1. Rank your threads in order from strongest to weakest.

2. Share your procedures and results with your classmates. Make a combined class data table. What general conclusions can be made about what makes a stronger thread?

3. How did the composition of the threads affect their breaking points? Give evidence to support your answer.

4. How is thread thickness related to strength?

5. Breaking strength, which is a force, should be expressed in newtons rather than a volume of water. We now need to calculate the total force (the weight of the water plus the plastic bottles plus the empty bucket) that caused each thread to break. The mass of 1 L of water is 1 kg. Obtain the mass of the empty bottles and bucket. You may recall that the weight of an object is the product of its mass times the acceleration of gravity ($9.8 \, m/s^2$): $W = m \times g$. Calculate the breaking-strength force for each of your thread samples.

6. Is the rank order of the strengths still the same?

Extend

Safety note: Wear safety goggles when working with steel cables.

1. Your teacher will provide you with some rope and cable samples. Compare the arrangement of the strands in each rope or cable. Are there any common designs? Make a sketch of each and note your observations.

Thread sample	Composition of thread	Observations with hand lens	Other observations
#1			
#2			

2. You will also be provided with a table comparing the strength (working load) of different types of rope. In the Explore stage, we used the breaking point to measure the strength of the threads. The working load of a rope is about 20% of its breaking point. This provides a margin of safety, so we use ropes and cables at a level far below where they may break. If the ¼-inch steel cable has a working load of 4,895 N, what would its breaking point likely be?

3. Examine the data table. How does the strength (as measured by working load) of the ropes and cables compare to their composition? Thickness? Arrangement of the strands?

4. In the Explore stage, you collected data using bottles of water to determine the breaking point. How many 0.5-L bottles of water would it take to break a cotton clothesline with a working load of 334 N (which is equivalent to 34.1 kg of water)? Show your work.

Evaluate

You have a ¼ in. twisted polypropylene clothesline that disintegrated quickly from being exposed to sunlight, which is a problem with this type of rope. Which of the other ¼ in. ropes listed in the table provided by your teacher would be at least as strong as the polypropylene rope? What other factors might you want to consider in your decision?

CHAPTER 14

SITTING AROUND DESIGNING CHAIRS

YOU MAY NOT be a couch potato, but most of us have a favorite chair—perhaps easily identified by its worn and possibly scruffy appearance. What is it that makes a favorite chair a favorite chair? Back and leg support? Arm rests? Most important, everything needs to fit you just right. Fitting one person just right, of course, may mean that another person would not find that chair especially comfortable—we are all different sizes.

For this reason, the engineers who design chairs must take into account the ergonomics of how chairs and humans (of different sizes) interact. *Ergonomics* is the "application of what we know about people, their abilities, characteristics, and limitations to the design of equipment they use, environments in which they function, and jobs they perform" (Human Factors and Ergonomics Society, n.d.). Each day, you likely sit in a variety of chairs—office chairs, stools at a counter, dining room chairs, folding chairs, possibly even a beanbag chair. Outdoors, you may sit on lawn chairs, park benches, or chaise lounges. Engineers must design these chairs, used for different purposes, with different criteria in mind. A *chair* is usually defined as an individual seat raised above the ground, usually with a back, and often with arms.

In this 5E learning-cycle lesson, using newspaper and tape, students design and build a chair that is capable of supporting their weight. Students must also consider the ergonomics of the chair they design so that it is dimensioned appropriately for typical middle-level

students. Teachers could modify this activity to meet the *Next Generation Science Standards*, which state that "[a] solution needs to be tested, and then modified on the basis of the test results, in order to improve it" (MS-ETS1-4; NGSS Lead States 2013). Because the main science content of this lesson involves balanced forces (the downward force of a person's weight must be supported by an equal and opposite force exerted by the chair to keep the person from breaking the chair, Goldilocks style), questions and tasks could be added to match (PS2.A: Forces and Motion).

Historical Information

Unless we are hiking or camping, most of us no longer sit on rocks. Humans have long been sitting on objects to get off the cold, damp floor or ground. Most early "chairs" were some type of bench or three-legged stool that focused on function. Our knowledge of ancient chairs mostly comes from art depictions or relics found in tombs, but it seems that people have been sitting on designed objects since the time of ancient civilizations.

Until the 1500s, only powerful or wealthy people sat in chairs; all others stood or, at best, sat on stools or benches. Thus, chairs were mostly designed to be large and highly decorated. We still see vestiges of this "chair power" in our language—the head of a committee or group is the chair, and the best violinist in an orchestra holds the first chair. Later, chairs became less grand and

CHAPTER 14

more practical so that during the Victorian Era, chairs were produced in sets for the first time (Kovel 2013).

While chairs likely seem commonplace and ordinary—essentially just a place for people to sit—they also represent the culture from which they come. One can deduce from early chairs much of what might be important to a culture, what materials were available, how the chairs were to be used, and their preferences in artistic expression. This importance of chairs to our culture is reinforced by Jessica Frazier (2012) of the Denver Art Museum, who notes that nearly all interior designers have made at least one chair: "Chairs combine form and function in a way that is easy for consumers to digest but incredibly difficult for designers to perfect inasmuch as they encompass many of the challenges of design—engineering, material choice, production method, style, and functionality—in one small package." Therefore, the topic of chairs makes an ideal choice for an everyday engineering activity.

Investigating How to Design Chairs (Teacher Background Information)

Materials

Have the students work in groups of four, if possible. For the Engage phase, you will need to locate three differently sized chairs. Ideally, use a tall stool, a typical classroom chair, and a child's chair. The goal is for two of the chairs to be the wrong size for middle-level students—so that the knees of the person seated are either much too high or too low (see Figure 14.1). While the amount required will vary depending on the type of chairs students build, for the Explore phase, each group will need a stack of newspaper about 30 cm (12 in.) high. You can ask students to bring in newspapers from home, or ask a local convenience store to save unsold papers for you (we were able to obtain a nearly unlimited supply in this manner). In addition, each group will also need a roll of masking or duct tape, a meter stick, and scissors. For the Extend phase, locate additional types and sizes of chairs, perhaps in the library, cafeteria,

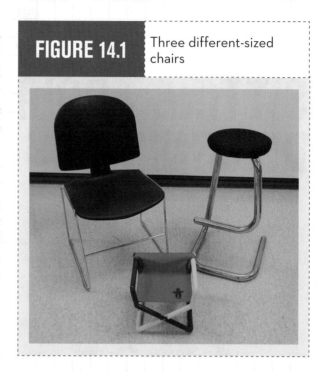

FIGURE 14.1 Three different-sized chairs

auditorium, or elsewhere. Alternatively, you may wish to have students conduct this part of the lesson at home. Finally, for the Evaluate phase, each group should have a small (30 × 30 cm or so) piece of corrugated cardboard for student purview. Old cardboard boxes are a good source for this sample. Have a fresh sample for each group so that students can manipulate freely to identify the characteristics.

Engage

To access students' prior knowledge, initiate a discussion about the chairs in which they are currently sitting, as well as some of their favorite places to sit when they are not at school. Ask students to think about what makes a chair comfortable. Present three widely different-sized chairs, and have each student predict which will be the most comfortable for them. Be sure that students state their rationale for the predictions.

Select one student to "test sit" each chair. Have students observe how well the size of each chair fits the size of the student, especially in regard to supporting the

FIGURE 14.2	A stool (left), a school chair (middle), and a child's chair (right)

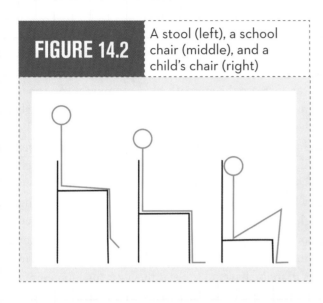

FIGURE 14.3	Taped rolls of paper to form a chair

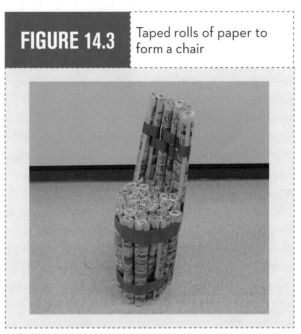

legs and feet. On a stool, you will likely notice that the knees are not supported by the seat, but angle downward, and the feet are not on the floor. Some stools have a footrest to support the feet. The child's chair also does not support the thighs and knees because of its very short height. Here, you should be able to notice that the knees of a student are angled upward (but the feet are flat on the floor). Neither of these sitting positions provides the leg and foot support necessary for comfort over an extended amount of time (see Figure 14.2). Finally, the regular classroom chair should be of the proper dimensions to provide support of the knee and thigh so that the feet rest on the floor and the thigh is parallel to the floor. Use the previous discussion to focus students on their task of designing an appropriately dimensioned chair that they will construct out of newspaper and tape. The main criteria for the activity are that the chair must be able to support the weight of one of the members of the group and be appropriately proportioned for the student for whom it is designed.

Explore

Discuss with the class the various ideas students suggest for constructing a chair out of newspaper. While there are several possible methods, a simple

option for strengthening the paper is to form it into tight rolls as shown in Figure 14.3. The rolls can be combined using the tape (see Figure 14.4). Each

FIGURE 14.4	Completed newspaper chair

group can now decide how they are going to use the newspaper to form a chair.

Students will need to determine the dimensions of their chairs. To do so, they might consider the dimensions of other chairs in the room or the distance from the back of the knee to the floor for students in their group. The distance from the back of the knee to the floor is known as the *popliteal height* and is widely used by chair designers. Each group will decide on which plan they wish to actually construct and test. Check each group's plan prior to their beginning construction for safety considerations and to allow for teacher feedback. Students should be free to try different construction methods with the newspaper. After construction, students should test their chairs and modify as necessary for comfort and strength. This iterative process is a crucial step in the engineering design of most products.

Plan on two class periods to complete this lesson. Materials could be stored in a labeled large trash bag for each group overnight. *Safety note:* Students should use caution with cutting tools and wear eye protection during the construction, testing, and disassembly of the chairs and wash the ink off their hands after working with the newsprint. Students should also use caution testing the different chairs to assure no one falls through a chair to the ground. They might want to "spot" each other as they conduct this testing. Most of the construction of the chairs will take place at students' tables; the testing will require only a minimal amount of floor space for each chair.

Explain

Have each group of students take a few minutes to share their chairs and then compare the different designs that were employed in their construction. Have students note the common characteristics of the chairs that were able to support a student without collapsing. If they are able, students should explain why certain designs were more successful than others. The teacher will need to help with these explanations when needed.

Students will find that a tight roll of paper is quite strong if they sit on the ends of the tubes positioned in a vertical position—that is, in compression. Students can increase the strength by combining many tubes together. They should discover that the rolls are much less strong if they are used to bear weight horizontally. In other words, if you sit on the sides of the rolls rather than the ends, the roll will be able to support much less weight. One sheet of paper is not very strong, but when rolled together, the load is distributed among the numerous sheets. The rolls are less strong when pushed or pulled from the sides. To compensate for this, engineers often will put three "tubes" together in a triangle shape, to transfer the horizontal forces into compressional ones. To this end, students may have noticed that tubes used on the surface of the seat, for example, may have buckled when under load. This can be compensated for by the use of triangular supports running from the bottom of the seats to the legs.

Also, the base of the chair in contact with the floor needs to be sufficiently broad to support the weight of the student. This is because the center of mass of the chair/student system must be within the base in contact with the floor in order for the system to be stable (and not tip over). Thus, a pedestal-type chair usually has legs at the bottom to have more contact with the floor (see Figure 14.5).

Extend

As students look at chairs, they may notice other ergonomic aspects of the chairs' designs such as chair back height and angle, seat angle, and leg design. For example, students may notice that many chairs have armrests. They might then measure the elbow rest height, which is the vertical distance from the seat to the armrest. The mean elbow rest height for adults in the United States is 23.4 cm (9.2 in.) for women and 24.1 cm (9.5 in.) for men (Georgia Institute of Technology, U.S. Department of Transportation 2000). The measure of human body dimensions in standard fixed postures is known as *anthropometry*, and such

FIGURE 14.5 Pedestal chair

FIGURE 14.6 Cardboard chair

published data are used by engineers to design a wide variety of products and human workspaces.

Several sets of standards have been developed for chairs and tables, especially those used in office or commercial settings. One set of standards is from BIFMA (the Business + Institutional Furniture Manufacturers Association; see Resources [p. 116]). They ensure the safety of the products and give guidelines for how they can be tested by both the manufacturer and the end user.

A few students may be familiar with lift chairs for people who need assistance getting out of a chair. Lift chairs use a mechanical system to push a person up onto their feet. It may be interesting to note some of the physics of getting out of a chair. In order for anyone to rise from a chair, they must lean forward to position their center of mass over their feet. It is impossible to get out of any chair otherwise. Usually we push down on the armrests to assist in this process. You may want to have students demonstrate this phenomenon.

Evaluate

Give each group of students a piece of corrugated cardboard so they can evaluate its characteristics as a building material for a chair. (See Figure 14.6 for an example of a corrugated cardboard chair.) Students now have more knowledge about chair construction that they can apply to this new challenge. For example, they will likely realize that cardboard is difficult to roll, which may lead them to stack it vertically rather than horizontally. Again, the vertical loading of the cardboard will be stronger than horizontally flexing it.

Conclusion

Every day, we use many simple, basic items that engineers have designed. When thinking of furniture making, many people think of a carpenter, an upholsterer, and perhaps an interior designer but may not realize that for the chair to be functional—stable, comfortable, and sufficiently strong—it must be engineered as well.

References

Frazier, J. 2012. A brief history of the chair in design. Denver, CO: Denver Art Museum. *http://denverartmuseum.org/article/staff-blogs/brief-history-chair-design*

Georgia Institute of Technology, U.S. Department of Transportation. 2000. ERGO TMC tools for user-centered design. Chapter 10: The workspace. Federal Highway Administration. *http://ergotmc.gtri.gatech.edu/dgt/Design_Guidelines/hndcha08.htm*

Human Factors and Ergonomics Society. n.d. Educational resources. *www.hfes.org/Web/EducationalResources/HFEdefinitionsmain.html*

NGSS Lead States. 2013. *Next Generation Science Standards: For states, by states.* Washington, DC: National Academies Press. *www.nextgenscience.org/next-generation-science-standards.*

Kovel, T. 2013. Have a seat: The history of chairs. *HeraldNet.* May 9. *www.heraldnet.com/article/20130509/LIVING03/705099995*

Resources

BIFMA website (Business + Institutional Furniture Manufacturers Association). *www.bifma.org/?page=standardoverview*

Gosnell, M. 2004. Everybody take a seat. *Smithsonian Magazine* 35 (4): 74–78. *www.smithsonianmag.com/people-places/everybody-take-a-seat-2386495/?no-ist=*

National Research Council (NRC). 2012. *A framework for K–12 science education: Practices, crosscutting concepts, and core ideas.* Washington, DC: National Academies Press.

ACTIVITY WORKSHEET 14.1   Investigating Chair Design

Engage

1. How would you describe the school chair you are sitting on right now? How does it compare to your favorite place to sit at home?

2. Your teacher has provided three different-sized chairs for a classmate to test. Predict which chair will be the best fit for your classmate. What are your reasons for your prediction?

3. One way to measure how comfortable a chair might be is how well a person's legs and feet are supported. For example, do the person's feet reach the floor? Are the person's thighs parallel to the floor and supported by the chair? Record your observations as one classmate sits in each of the three chairs.

4. Which chair was the best fit for the support of the student's legs and feet?

5. Keeping this information in mind, you will design and build a chair made of newspapers and tape.

Explore

1. There are two criteria for this design challenge. One is to design a chair that has the proper dimensions for a selected student in your group. The second is that it must readily support that student's weight for a period of time—at least 10 seconds. Brainstorm some ideas of how to build a chair out of only newspaper and tape.

2. Share your groups' ideas with the rest of the class and explain why you think it is a good design. After you have discussed the ideas, decide on a design for your group's chair.

3. Now that you have an idea for the construction of your chair, you need to decide its dimensions. For example, you will need to take some measurements of your model student to determine how high the seat should be above the floor.

4. Make a detailed sketch of your design that includes dimensions. Have your teacher approve your design before you start to build.

5. Use the newspaper and tape to construct your chair. Test it for proper fit, and see whether it can support a middle-level student and make modifications if needed.

Explain

1. Share your completed chair with the rest of the class. How well does your chair fit the student for whom it was designed? Discuss possible ways to improve your group's design.

2. How are all of the newspaper chairs in your class similar and different? Think of the chair as a system. What are some common characteristics of the chairs that were able to support a student?

3. How is it that something as weak as newspaper could be made strong enough to support a student?

Extend

1. What are some other dimensions or characteristics that should be considered when designing a chair?

2. Locate several different types of chairs at home or at school and measure critical dimensions and note significant characteristics. If possible, take photos or make a sketch to record and be able to share your data.

3. How does the intended use of a chair affect its design?

Evaluate

Suppose you wanted to design another chair, but had a supply of corrugated cardboard instead of newspaper. How would you alter your design to accommodate a different material? Make a sketch of your new design and write a description of how you would go about constructing and testing the new chair.

INDEX

Page numbers printed in **boldface** type refer to figures or tables.